Black and White

Volume 10

Smart Things Begin With Griddlers.net

Griddlers Logic Puzzles: Black and White (Volume 10)

Published by: Griddlers.net
a division of A.A.H.R. Offset Maor Ltd

Author: Griddlers Team
Compiler: Rastislav Rehák
Cover design: Elad Maor
Contributors: Adrianko, Agrippina, alibaba, anawa74, Andreasss, Ardna, bart88, carm53, carrben12, eleninikolaoy, elimaor, esra, fineke, fordpre, Glucklich, griddlock, HelleRys, hero692ttr, hibrahimozer, Hoborg, inro, ireneh, janiebugsmommy, julesc, komplexyapi, ld5, Lupos, maarten001, macro, maristone, maycanatan, muniaelfik, mustafademirbas, oko, op1, Oskar, Ra100, rapa100, rehacik, sandyeggan, shadow2097, shybobcat, starch, stomlins123, stumpy, thorny77032, TNT, tono, tulipe, twoznia, tygerbug, xiayu, xxLadyJxx, yahoo, zz89

ISBN: 978-9657679098

More information:
Email – team@griddlers.net
Website – http://www.griddlers.net

Definition

Griddlers, also known as Paint by Numbers or Nonograms, are picture logic puzzles in which cells in a grid have to be colored or left blank according to numbers given at the side of the grid to reveal a hidden picture. In this puzzle type, the numbers measure how many unbroken lines of filled-in squares there are in any given row or column. For example, a clue of "4 8 3" would mean there are sets of four, eight, and three filled squares, in that order, with at least one blank square between successive groups.

These puzzles are often black and white but can also have some colors. If they are colored, the number clues will also be colored in order to indicate the color of the squares. Two differently colored numbers may or may not have a space in between them. For example, a black four followed by a red two could mean four black spaces, some empty spaces, and two red spaces, or it could simply mean four black spaces followed immediately by two red ones.

There are no theoretical limits on the size of a nonogram, and they are also not restricted to square layouts. (From Wikipedia, the free encyclopedia)

Basic Rules

Each clue indicates a group of contiguous squares of like color.

Between each group there is at least one empty square.

The clues are already in the correct sequence.

Groups of different colors or different triangles may or may not have empty squares between them. Triangle without a number counts as 1.

Step by Step Example

Following example will show you how to solve easy puzzle. Mark empty squares by dot. We suggest to use marking especially for difficult puzzles.

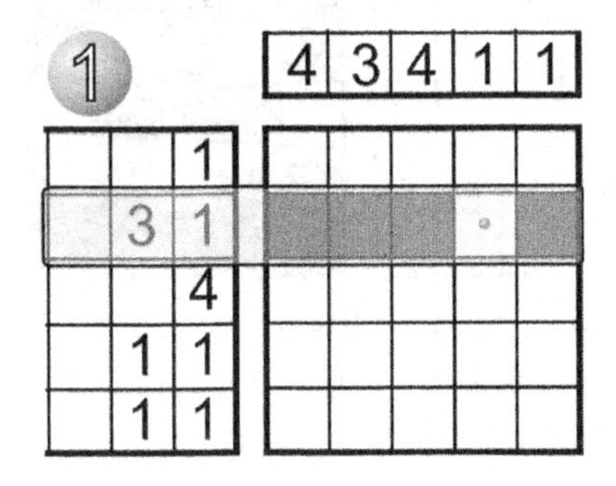

Row 2: Clues (3,1), with 1 empty square (background color) between them, add up to the 5 available squares.

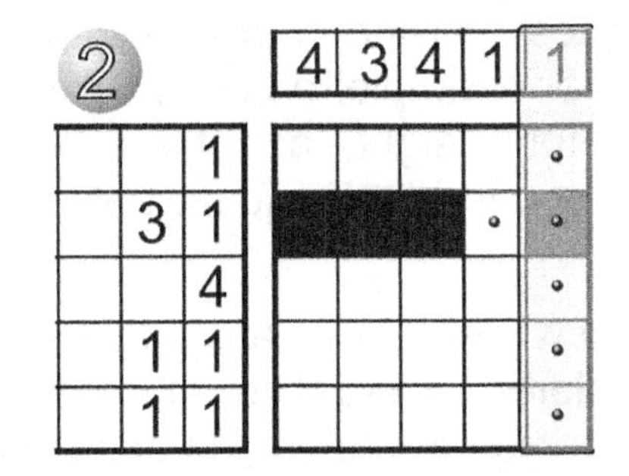

Column 5: Clue 1 is already on the grid. We can fill in the rest with background color.

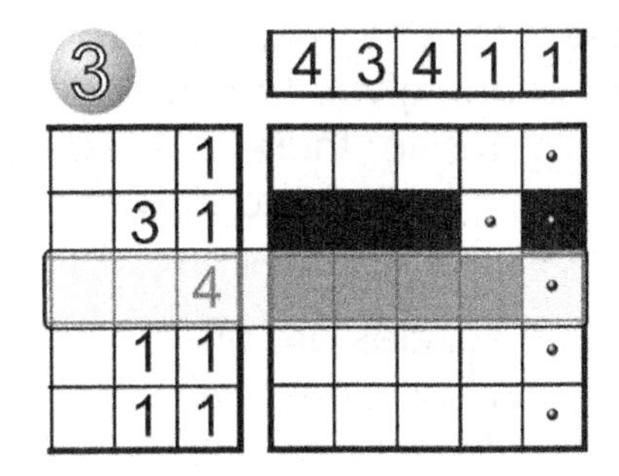

Row 3: There are only 4 squares left to place clue 4 on the grid.

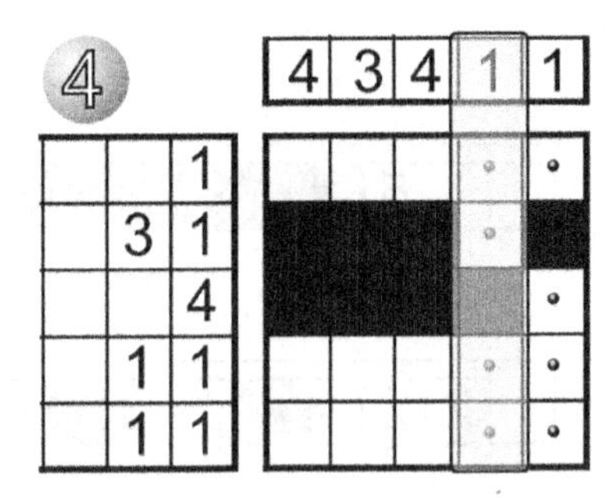

Column 4: Clue 1 is already on the grid. We can fill in the rest with background color.

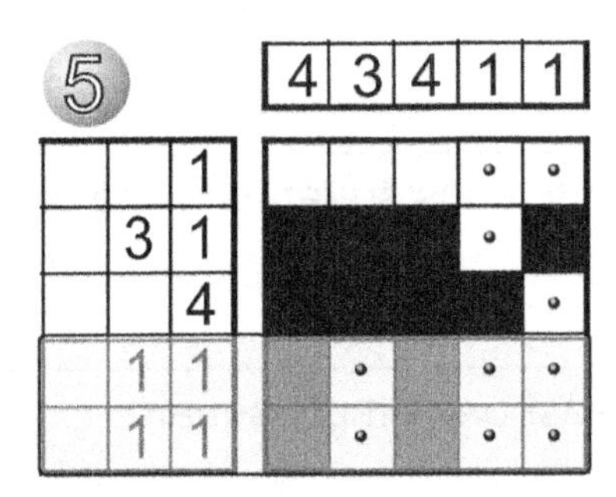

Row 4/5: Clues (1,1), with 1 empty square between them, add up to 3 available squares.

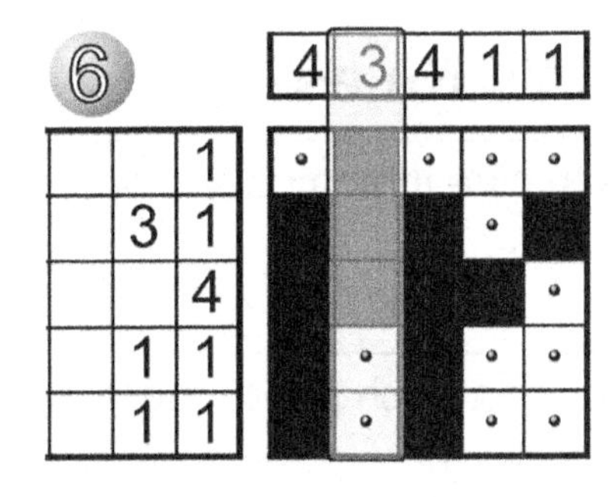

Column 2: There is only 1 square left to complete clue 3 and finish the puzzle.

Names

Nonograms are also known by many other names, including Paint by Numbers, Griddlers, Pic-a-Pix, Picross, PrismaPixels, Pixel Puzzles, Crucipixel, Edel, FigurePic, Hanjie, HeroGlyphix, Illust-Logic, Japanese Crosswords, Japanese Puzzles, Kare Karala!, Logic Art, Logic Square, Logicolor, Logik-Puzzles, Logimage, Oekaki Logic, Oekaki-Mate, Paint Logic, Picture Logic, Tsunamii, Paint by Sudoku and Binary Coloring Books.
(From Wikipedia, the free encyclopedia)

1

Column clues:

								1	1					1	2									
			5			2	2	2	2	4			1	4	2	2								
1	1		1			5	3	1	1	1			2	1	1	1		3	2	1	1			
1	2		1	3	3	2	1	1	1	3			1	2	2	2	3	2	6	1	1		1	
2	1	1	2	1	1	2	2	4	4	2	13		2	1	1	1	1	6	2	1	1	1	2	1
2	1	1	1	3	4	1	3	1	2	3	2	13	1	1	2	1	1	1	1	1	2	1	1	1
1	2	3	4	2	1	1	2	1	1	1	4	3	1	1	2	3	1	1	2	1	1	1	1	1

Row clues:

				6
			3	6
		3	4	2
		3	6	2
	2	2	4	2
1	2	2	2	2
1	2	2	2	1
	1	1	2	1
1	2	1	2	1
15	1	2	1	2
		2	1	1
		2	9	2
3	2	7	2	3
	2	3	4	2
1	2	3	1	2
		2	10	2
				18
2	1	3	2	2
1	3	2	2	3
			2	3
		1	4	3
		2	1	3
	3	2	2	2
	4	4	3	3

25x25 97 xiayu

2

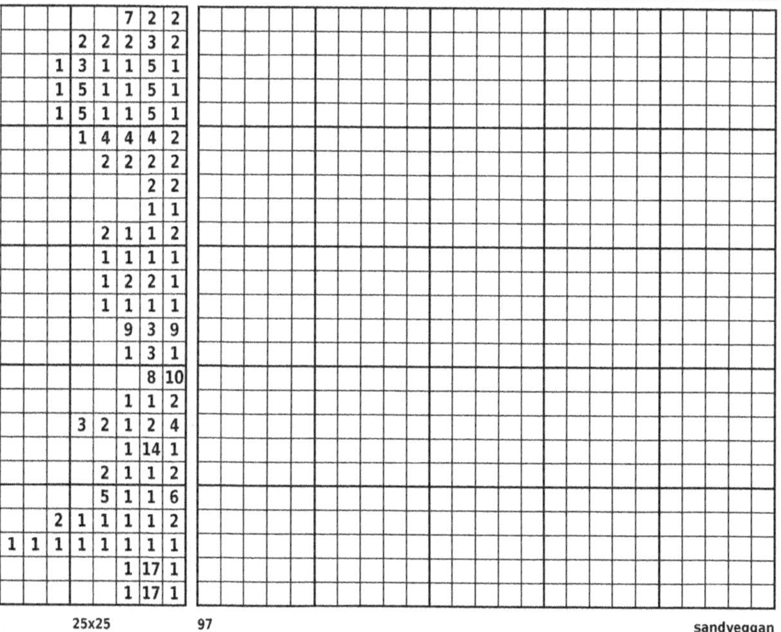

Column clues:

					1																			
					2	1		1												3	1			
				2	3	4	1	5								1			5	3	2	2		
				2	1	5	5	1	1						2	4	6	6	5	1	1	2		
			6	1	1	1	3	1	3	2		1		5	1	1	1	3	1	1	1	1	5	
		1	1	1	1	2	1	4	4	2	5	2	1	4	1	4	2	1	2	1	1	1	1	
	1	1	1	3	2	1	1	1	2	1	2	3	1	1	2	1	1	1	1	1	2	3	1	1
1	1	3	1	1	3	2	2	2	5	2	7	2	2	7	2	2	5	2	2	2	3	1	1	3

Row clues:

					7	2	2
			2	2	2	3	2
		1	3	1	1	5	1
		1	5	1	1	5	1
		1	5	1	1	5	1
			1	4	4	4	2
				2	2	2	2
						2	2
						1	1
				2	1	1	2
				1	1	1	1
				1	2	2	1
				1	1	1	1
					9	3	9
					1	3	1
						8	10
					1	1	2
			3	2	1	2	4
					1	14	1
				2	1	1	2
				5	1	1	6
		2	1	1	1	1	2
1	1	1	1	1	1	1	1
					1	17	1
					1	17	1

25x25 97 sandyeggan

3

Column clues:

												1						1	2	2									
							3				2	2		1	6	6	1	2	1	3		2							
				2			2		4	3	4	2	1	2	4	2	2	1	3	1	4	1							
				2		2	3	6	3	3	1	1	5	2	2	2	5	2	3	1	3	1							
1	2	1	2	4		1	4	4	3	1	2	1	1	1	1	2	1	1	5	4	4	2	3			1	1		
2	2	2	2	2	18	5	3	6	3	3	3	2	2	1	1	8	2	6	1	6	3	12	2	3	3	1	2	1	1
2	3	4	2	2	3	3	1	8	3	1	1	3	2	3	3	3	4	2	3	1	6	1	2	5	5	3	2	2	1

Row clues:

						1	2	1
						2	2	2
						2	3	2
						3	2	4
						4	3	4
					2	10	3	1
					3	3	5	3
				3	2	6	1	4
						4	11	4
1	1	2	1	1	2	1	4	1
					5	11	4	6
				6	3	4	4	7
				1	6	1	3	1
					2	1	3	1
					2	1	3	1
						2	5	5
					2	3	1	2
						1	7	7
				2	2	3	2	3
				2	2	1	2	1
					2	2	1	4
				2	1	1	2	2
					2	3	3	5
				2	2	1	3	1
				5	1	2	1	3
					2	2	2	5
					2	2	8	2
				2	1	2	6	3
				3	2	2	6	1
				3	2	3	1	5

30x30

97

starch

4

Column clues:

																								3	5				
																								2	2	5		3	3
																		4				3	3	2	1	1		2	1
										3			2			3	3	2			4	1	2	1	1	1	4	1	1
										1	3		3		3	2	1	2	6		1	1	3	1	1	1	2	1	1
1										3	2	3	1	3	3	1	1	3	2	6	1	1	1	1	1	3	1	1	1
1							3		5	2	1	2	2	3	1	1	4	3	1	2	1	1	1	1	1	1	1	3	3
1							1		1	1	1	3	1	2	1	1	1	1	3	1	1	1	1	1	1	1	3	1	1
1					6	5	1	3	2	1	1	6	2	3	7	7	3	1	1	1	1	1	1	1	1	1	1	1	1
1			2	4	1	2	1	1	2	1	3	1	1	3	1	1	1	1	1	7	3	1	1	1	1	1	5	1	1
1	1	1	1	3	3	2	2	5	2	1	3	2	4	2	1	3	2	1	2	1	1	2	4	2	1	2	1	1	1

Row clues:

								7
								13
							1	22
							9	10
							7	9
				1	4	1	3	4
						4	9	5
					1	6	2	2
				1	1	3	1	1
							2	2
				1	4	1	1	4
				1	2	3	1	1
	1	1	1	2	2	3	1	1
1	2	1	2	2	1	1	1	1
	5	1	2	7	1	1	1	1
2	1	3	3	1	1	1	1	1
	1	2	3	1	1	1	1	1
	2	3	3	1	1	1	3	1
		2	5	1	1	1	1	3
			2	9	1	1	1	3
		5	3	3	1	1	1	2
2	1	1	3	3	1	1	1	1
1	1	1	5	1	1	1	1	1
2	1	3	1	3	1	1	1	1
			6	1	1	3	1	3
	3	1	1	3	1	1	1	1
			2	1	1	1	3	3
			4	1	3	1	1	1
					5	2	1	1
							4	1

30x30

97

thorny77032

5

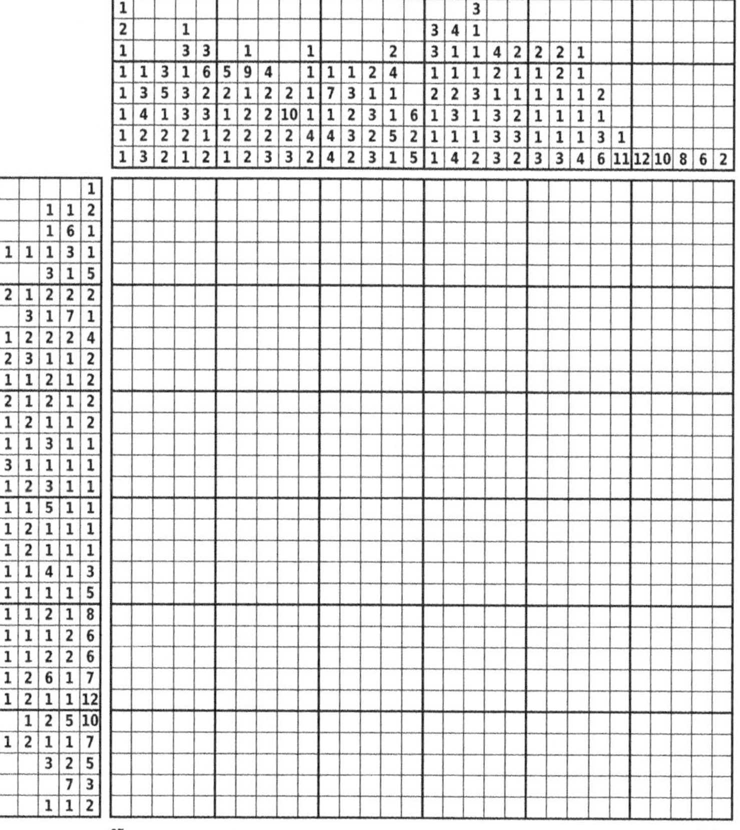

1 2 1 1 1 1 1 1
1 3 4 2 3
3 5 1 2 2
1 3 1 3 3 2 1
3 6 2 3 1 2
5 2 1 2 1
1 9 1 2 2 2
4 2 2 2 3
2 10 2 3
1 1 1 1 4 2
1 7 1 4 4
1 3 2 3 2
2 1 3 2 3
2 4 1 1 5 1
6 2 5
3 3 1 2 1 1 1
4 1 1 2 3 1 4
1 3 1 1 1 3 1 1 2
4 2 1 3 3 3
2 1 1 2 3 2
2 1 1 1 1 3
2 2 1 1 1 3
1 1 1 1 1 4
2 1 3 6
1 11
12
10
8
6
2
1
1 1 2
1 6 1
1 1 1 3 1
3 1 5
2 1 2 2 2
3 1 7 1
1 2 2 2 4
3 2 3 1 1 2
2 2 1 1 2 1 2
1 2 1 2 1 2 1 2
2 2 1 1 1 2 1 1 2
2 1 1 1 1 3 1 1
4 1 1 3 1 1 1 1
1 1 1 1 1 1 2 3 1 1
1 1 1 1 1 1 5 1 1
3 1 3 1 2 1 1 1
1 1 1 1 1 1 1 1 2 1 1 1
1 1 1 1 1 1 1 4 1 3
1 1 1 1 1 1 1 1 5
1 3 1 1 1 1 2 1 8
1 1 1 1 1 1 1 2 6
2 2 1 1 3 1 1 2 2 6
1 1 1 1 2 6 1 7
1 2 1 1 12
1 2 5 10
1 2 1 1 7
3 2 5
7 3
1 1 2
30x30
97
Oskar

6

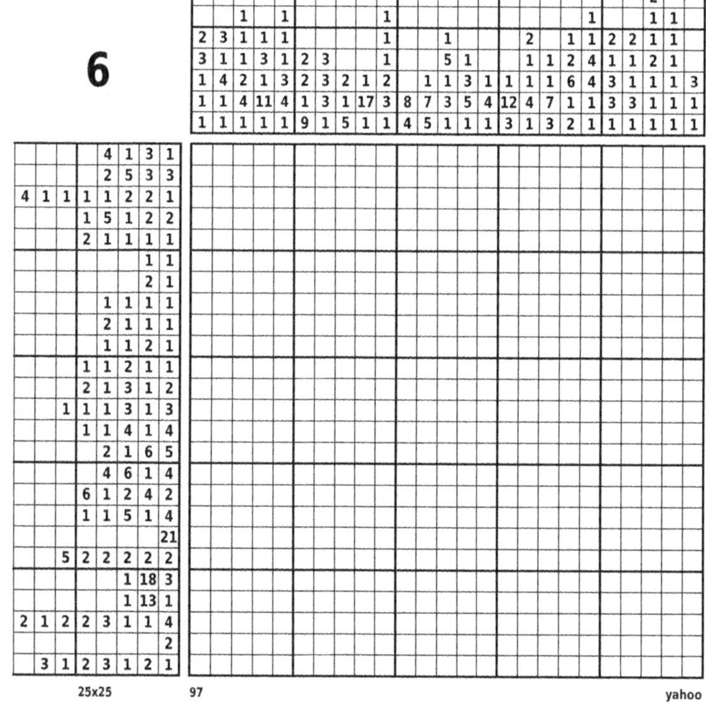

2 3 1 1 1
3 1 4 1 1
1 1 1 2 4 1
1 3 1 11 1
1 1 1 3 4 1
2 2 1 9
3 3 3 1
2 1 5
1 17 1
1 1 1 2 3 1
8 4
1 7 5
1 5 1 3 1
1 3 5 1
1 4 1
1 12 3
2 1 1 4 1
1 1 7 3
1 2 6 1 2
1 1 4 4 1 1
2 1 3 3 1
2 1 1 3 1
2 1 1 2 1 1 1
1 1 1 1 1 1
3 1 1
4 1 3 1
2 5 3 3
4 1 1 1 1 2 2 1
1 5 1 2 2
2 1 1 1 1
1 1
2 1
1 1 1 1
2 1 1 1
1 1 2 1
1 1 2 1 1
2 1 3 1 2
1 1 1 3 1 3
1 1 4 1 4
2 1 6 5
4 6 1 4
6 1 2 4 2
1 1 5 1 4
21
5 2 2 2 2 2
1 18 3
1 13 1
2 1 2 2 3 1 1 4
2
3 1 2 3 1 2 1
25x25
97
yahoo

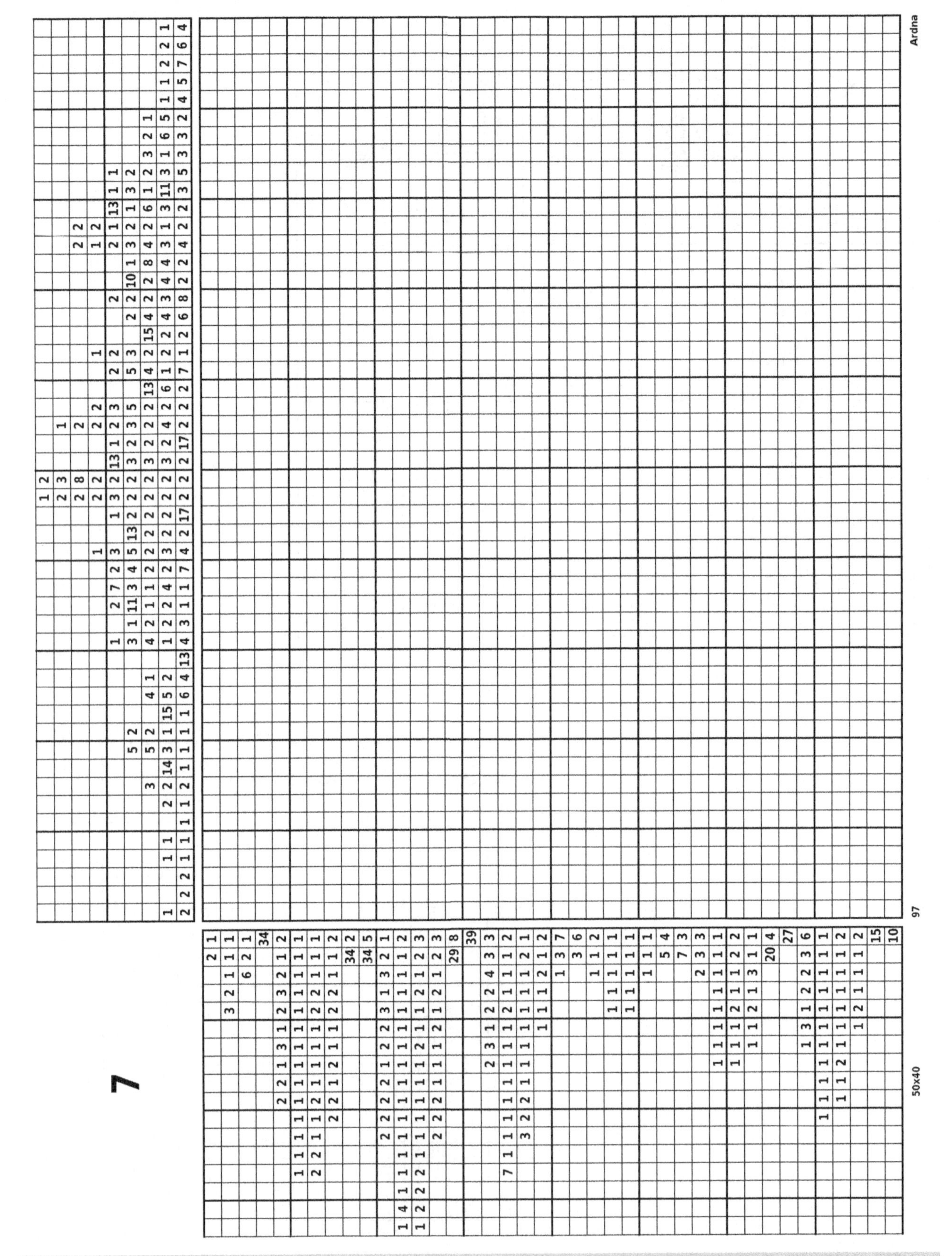
7
50x40
Ardna
97

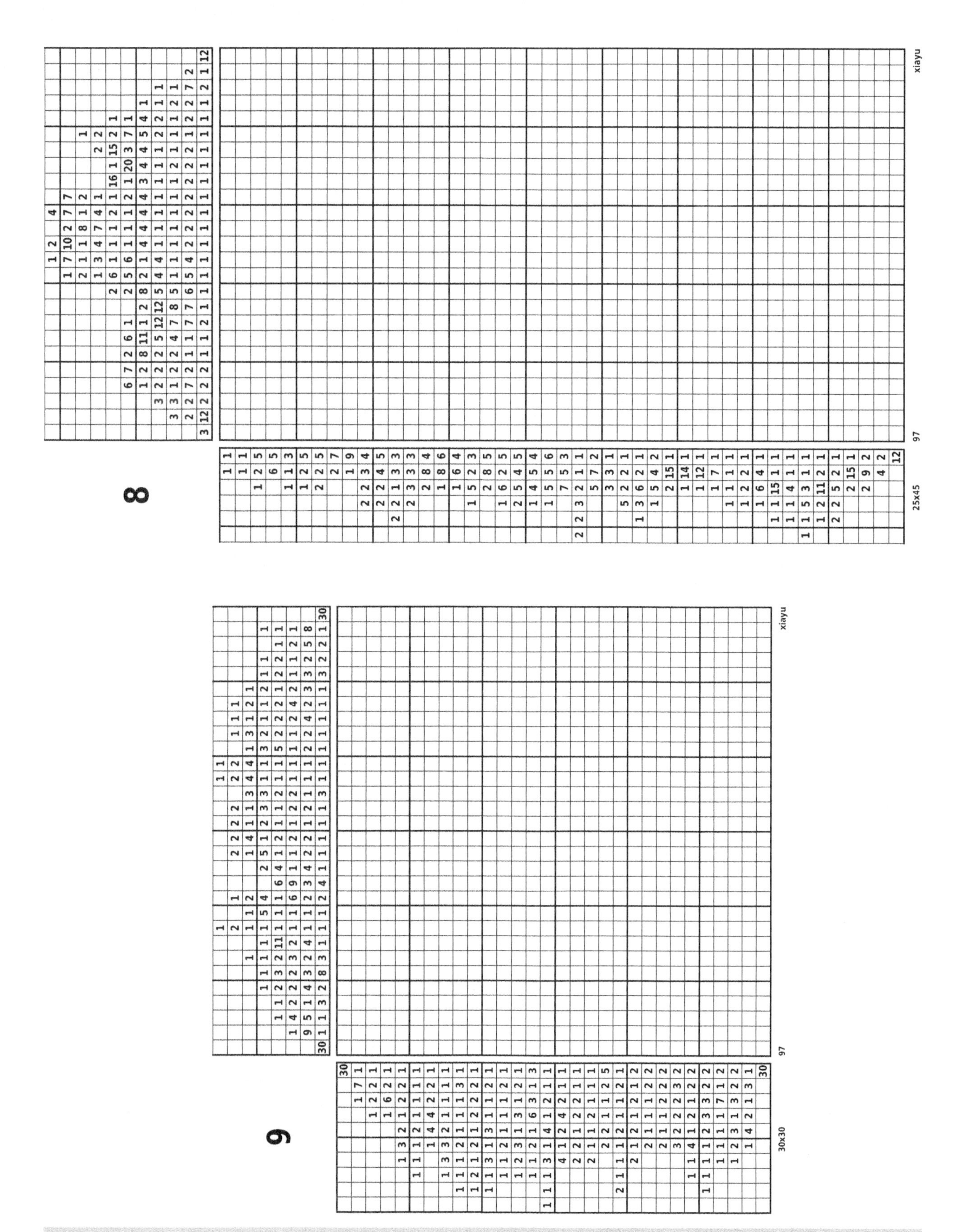

10

																														1														
																														1														
												1	1					1										1	1	2		1												
						1	1	1		1	1	1	1	1	1		1	1								1	1	1	1	1	1	1												
						1	1	1	1	1	1	3	1	1	1		1	1	1					1	1	1	1	5	4	1	1	1	1											
						18	17	17	1	6	5	3	1	1	1	1	1	2	1	1				1	1	16	13	4	2	1	1	1	1	1										
					1	2	2	1	16	7	5	1	1	1	1	1	1	3	1	1	1		1	7	16	1	1	1	1	1	1	1	1	1										
					1	3	2	1	1	1	1	1	1	1	1	1	1	2	13	13	1	1	1	10	8	4	1	1	1	1	1	4	5	14	1									
			1	1	14	3	2	1	1	1	1	1	2	4	6	1	1	15	15	1	1	26	6	1	1	1	1	1	1	3	2	9	14	13	1	1								
			1	5	16	1	2	1	1	1	2	1	3	5	6	15	15	1	1	2	11	1	1	2	1	1	3	1	1	1	10	1	1	2	1	33	1							
			35	1	2	1	1	2	1	1	1	2	1	1	1	2	1	1	1	1	2	1	1	1	2	1	1	2	1	1	1	2	1	1	2	1	2	37	1	1				
	44	1	1	1	1	2	1	1	2	1	1	1	2	1	1	1	2	2	1	1	2	1	1	2	1	1	2	1	1	2	1	1	2	1	1	2	1	1	1	2	3	1	2	
5	1	3	1	1	1	1	1	1	1	1	1	1	1	1	2	1	1	1	1	1	2	1	1	2	1	1	2	1	1	2	1	1	2	1	1	2	1	1	2	1	1	2	1	4

								23
							15	1
						1	21	1
				1	7	1	1	1
		1	6	5	1	10	1	1
1	1	7	1	1	6	1	1	1
1	1	7	1	1	5	1	1	1
1	1	7	1	1	5	1	1	1
1	1	6	2	1	4	1	1	1
1	1	6	2	1	3	1	1	1
1	1	6	3	1	3	1	1	1
1	1	5	4	1	6	1	1	1
1	1	9	2	1	5	1	1	1
1	1	8	2	1	5	1	1	1
1	1	8	2	1	5	2	1	1
1	1	7	3	1	4	3	1	1
1	1	7	3	1	4	2	1	1
1	1	6	3	1	3	2	1	1
1	1	6	1	1	3	6	1	1
	1	1	5	2	1	4	1	1
	1	1	5	2	1	7	1	1
1	1	4	5	1	4	3	1	1
	1	1	8	4	2	4	1	1
1	1	2	7	1	3	5	1	1
1	1	6	4	1	4	2	1	1
1	1	3	5	1	3	3	1	1
1	1	2	5	1	3	3	1	1
1	1	1	6	1	2	4	1	1
1	1	4	6	1	2	4	1	1
1	1	3	7	1	1	5	1	1
1	1	2	8	1	2	4	1	1
1	1	1	4	1	1	4	1	1
1	1	1	5	1	1	5	1	1
1	1	1	5	1	1	4	1	1
1	1	4	6	1	1	4	3	1
	1	1	3	6	1	5	4	2
		1	1	2	7	1	5	7
	1	1	1	6	1	4	4	1
		1	1	1	4	5	5	5
	1	1	1	5	5	4	5	2
		1	1	3	5	5	4	1
			1	1	5	5	4	3
			1	1	4	5	4	4
				1	3	5	4	4
					1	4	4	4
					1	5	4	4
						3	4	4
					1	3	4	4
						1	7	4
							1	4

45x50

97

xxLadyJ xx

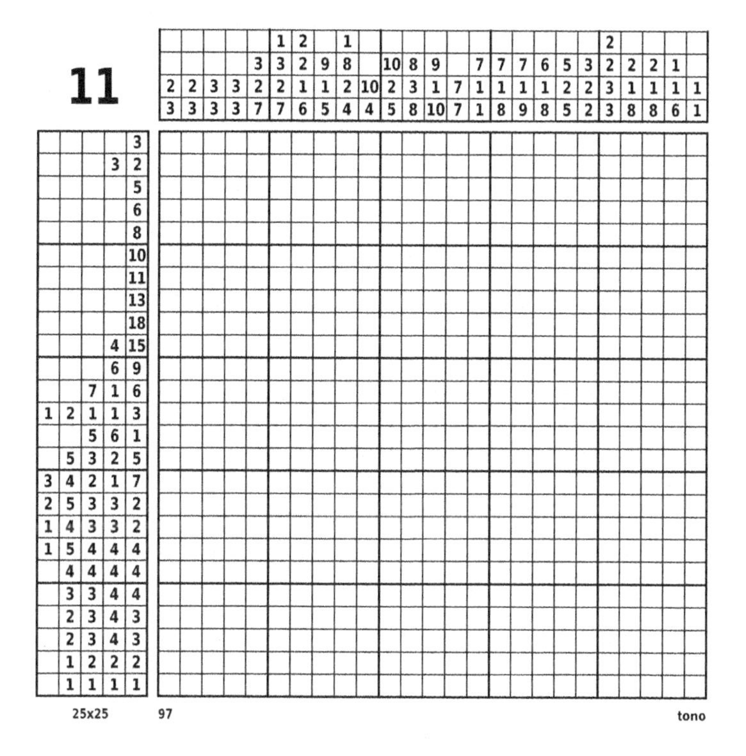
11
25x25
97
tono

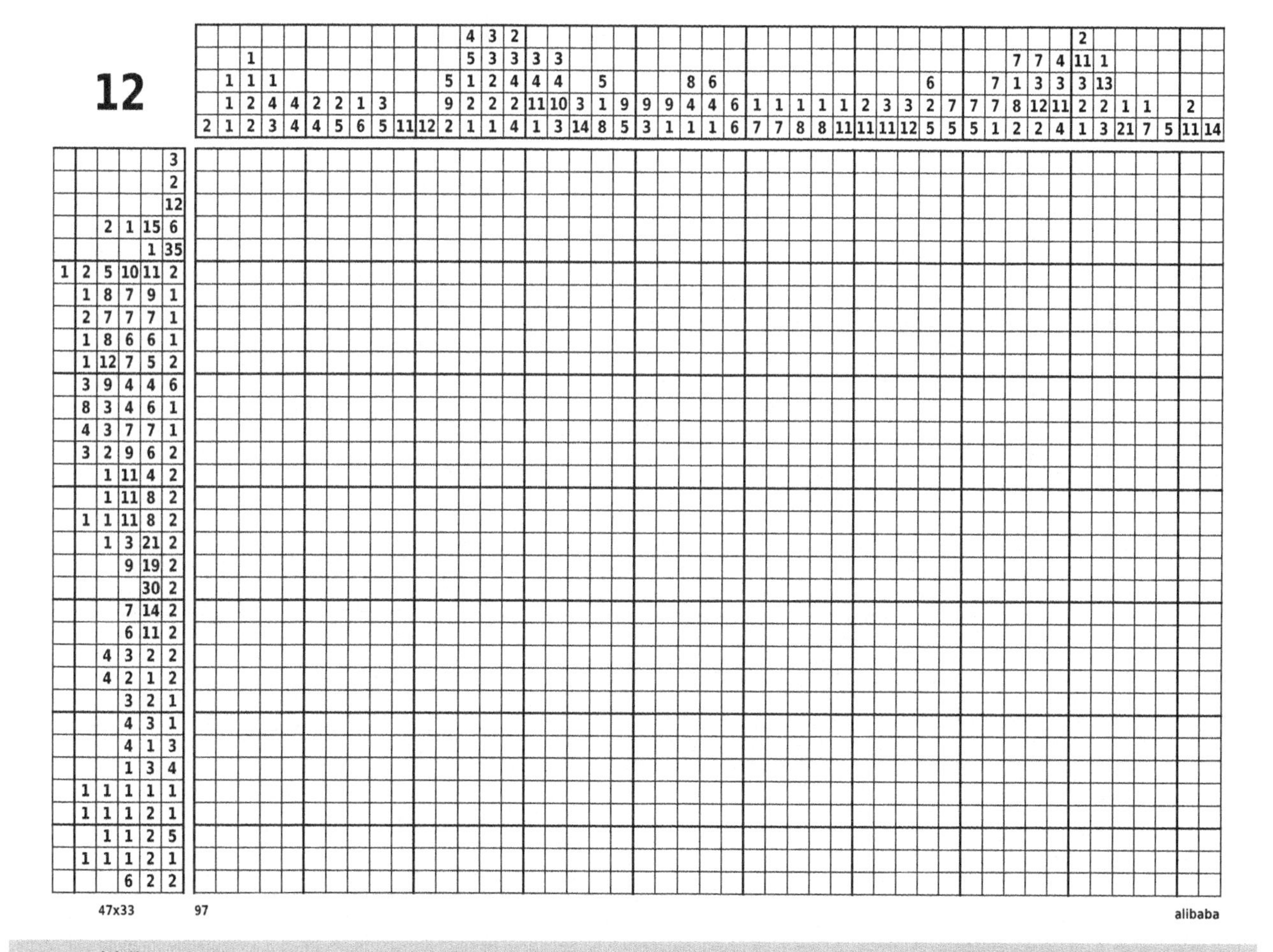
12
47x33
97
alibaba

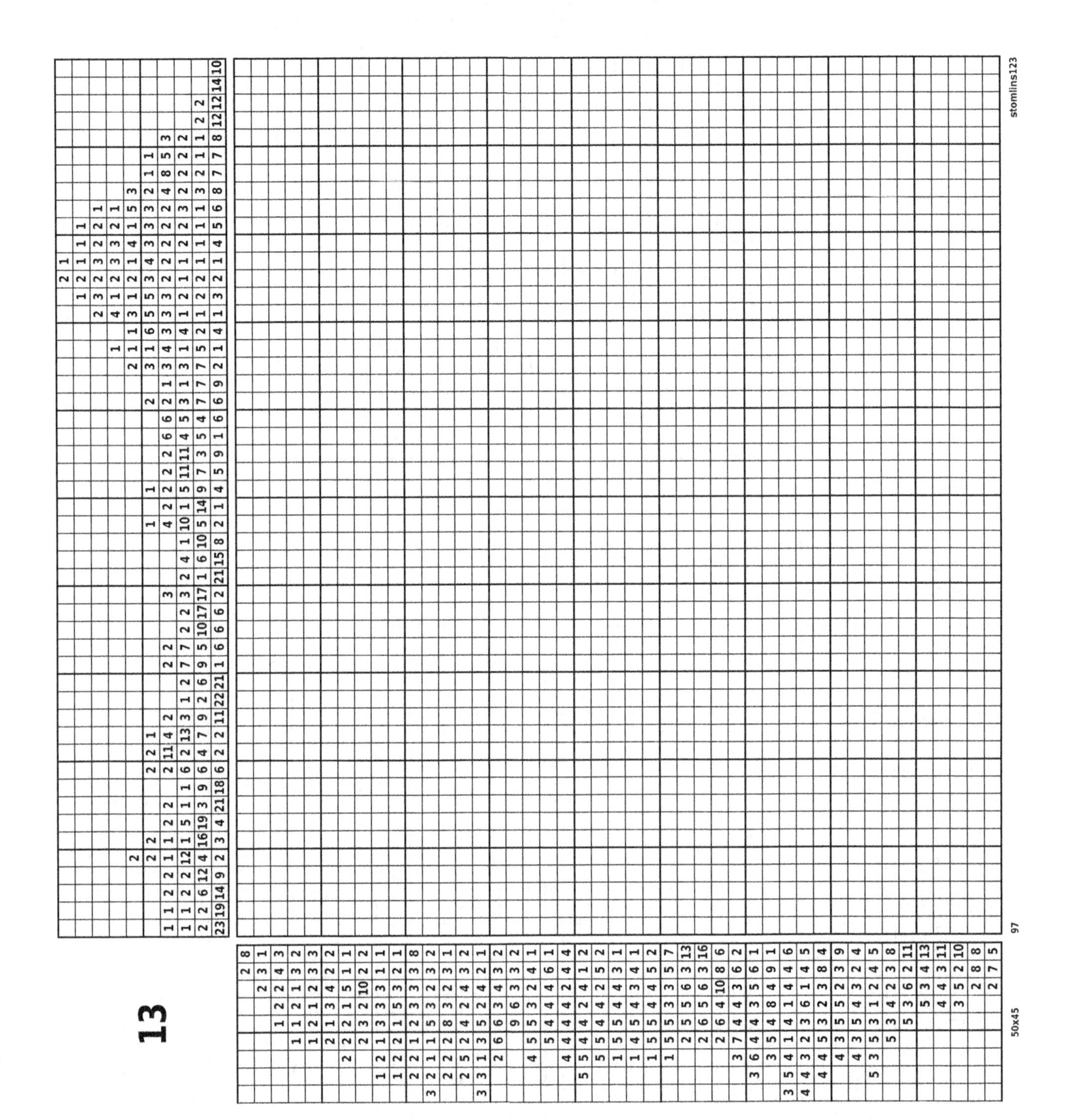
13
50x45
97
stomlins123

14

			3		3		3			
		2	1	1	1	1	1	2		
	3	3	1	10	1	10	1	3	3	
	2	3	1	12	1	12	1	3	2	
5	2	1	1	1	2	1	1	1	2	5

					3
			1	1	1
			1	1	1
				1	1
				1	1
				1	1
				1	1
				1	1
				1	1
				1	1
				1	1
				1	1
				1	1
					7
				1	1
		1	1	1	1
		1	1	1	1
		2	1	1	2
		1	1	1	1
		2	1	1	2
		1	1	1	1
			1	5	1
1	1	1	1	1	1
1	1	1	1	1	1
1	1	1	1	1	1
			1	5	1
		1	1	1	1
				2	2
			1	1	1
					3

11x30 98 twoznia

15

						2	2		
1	1	1	1	1		1	1		1
1	1	1	1	1		1	1	6	1
1	1	1	1	1		1	2	1	1
1	1	3	3	3	18	1	1	1	1
1	1	1	1	1	1	1	1	1	1

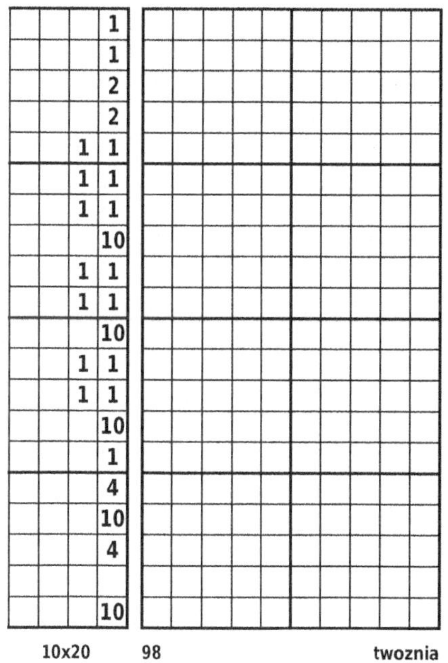

	1
	1
	2
	2
1	1
1	1
1	1
	10
1	1
1	1
	10
1	1
1	1
	10
	1
	4
	10
	4
	10

10x20 98 twoznia

16

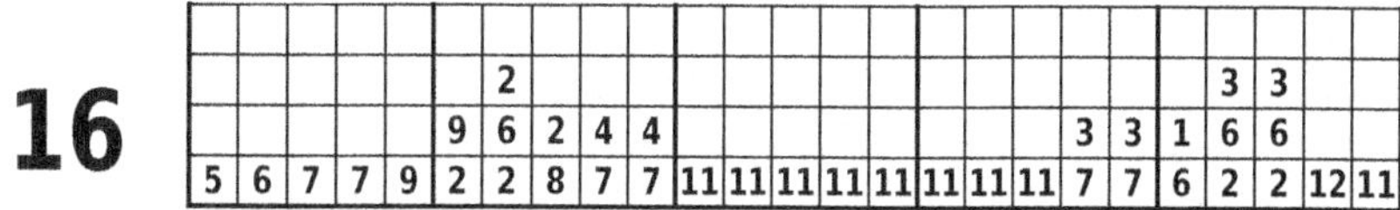

						2															3	3		
					9	6	2	4	4									3	3	1	6	6		
5	6	7	7	9	2	2	8	7	7	11	11	11	11	11	11	11	11	7	7	6	2	2	12	11

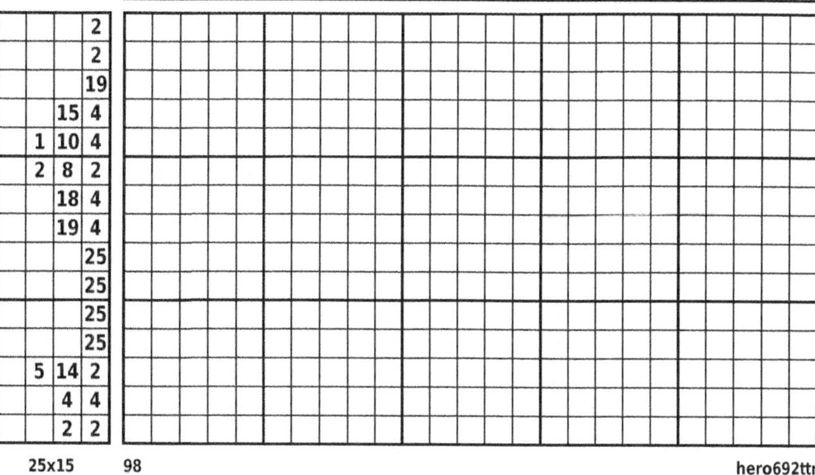

		2
		2
		19
	15	4
1	10	4
2	8	2
	18	4
	19	4
		25
		25
		25
		25
5	14	2
	4	4
	2	2

25x15 98 hero692ttr

17

Column clues:

Column	Clue
1	4 2
2	6 2
3	7 1 3
4	5 2 2
5	1 1 2 2
6	2 1 2 2
7	3 1 2 2
8	2 1 2 2
9	3 1 2 2
10	2 5 2 2
11	2 2 5 2 2
12	3 1 3 1 2 2 2 1
13	2 1 7 2 2 2 2
14	5 8 1 2 2 2
15	2 7 1 2 2 2
16	11 2 1 2
17	1 12 1 2
18	2 13 7
19	2 10 8
20	16 3
21	2 14
22	3 13
23	2 22
24	32
25	5 23
26	23
27	27
28	15 23
29	12 23
30	2 2 21
31	2 20
32	2 20
33	2 20
34	2 20
35	2 20
36	2 19
37	2 19
38	2 17
39	2 17
40	2 15
41	2 13
42	2 11
43	3 7
44	2
45	2

Row clues:

Row	Clue
1	3
2	4
3	1 6
4	2 5
5	3 3 2 2
6	3 1 2 3
7	2 1 2 2
8	2 1 2 2
9	3 5 1 2
10	2 2 2 1 2
11	2 1 3 3 2 2
12	1 1 5 2 2 2
13	7 6 2 2 2
14	2 6 1 2 2
15	8 1 2 2
16	1 6 1 2 2
17	1 6 1 2 2
18	2 11 2
19	2 10 2 3
20	16 4
21	3 11 4
22	4 25
23	34
24	22
25	13
26	21
27	5 19
28	12 20
29	10 20
30	3 21
31	2 20
32	2 20
33	2 21
34	2 21
35	2 21
36	2 21
37	2 21
38	2 20
39	2 20
40	2 18
41	2 18
42	2 16
43	2 14
44	2 12
45	2 8

45x45 97 xxLadyJ xx

18

							3	3				5	4							4				3										
							2	2	3		4	2	1	4			5	6	3	1		3	6	3				4	4					
					2		2	2	2	2	2	2	3	1	3	7	1	1	4	3	11	1	3	3	4			2	1	3				
				2	4	9	3	4	1	2	1	1	3	3	2	3	3	3	3	1	4	3	11	1	2	5	7	2	2	1	3			
				4	4	3	1	1	3	4	3	4	3	6	2	1	7	7	7	6	4	13	1	10	3	3	2	4	1	3	3	3		
			2	1	2	3	2	2	4	3	5	1	5	9	18	16	13	14	13	12	12	7	3	2	12	11	10	4	7	4	2	4	2	1
2	4	4	6	6	8	9	10	11	12	13	14	12	10	8	6	5	5	4	4	4	4	4	5	6	10	10	10	8	6	5	3	1	1	1

							9
							13
							15
				6	2	1	6
				7	1	2	5
		3	4	1	3	1	3
1	4	1	1	2	2	2	2
				3	2	7	6
				4	5	6	2
				3	4	16	1
				2	1	20	1
					2	10	3
			5	4	2	6	2
	3	3	3	4	4	3	1
				3	6	3	11
				4	4	3	10
			6	3	4	6	2
				1	3	7	8
					3	4	17
						8	18
				6	3	2	11
				4	2	5	8
					4	6	5
						8	13
					2	3	10
						4	10
					2	2	9
						3	9
						6	9
						8	11
					11	8	4
					13	9	4
					15	8	6
					15	7	7
					15	5	9
						15	9
							26
							24
							21
							17

35x40

98

maristone

19

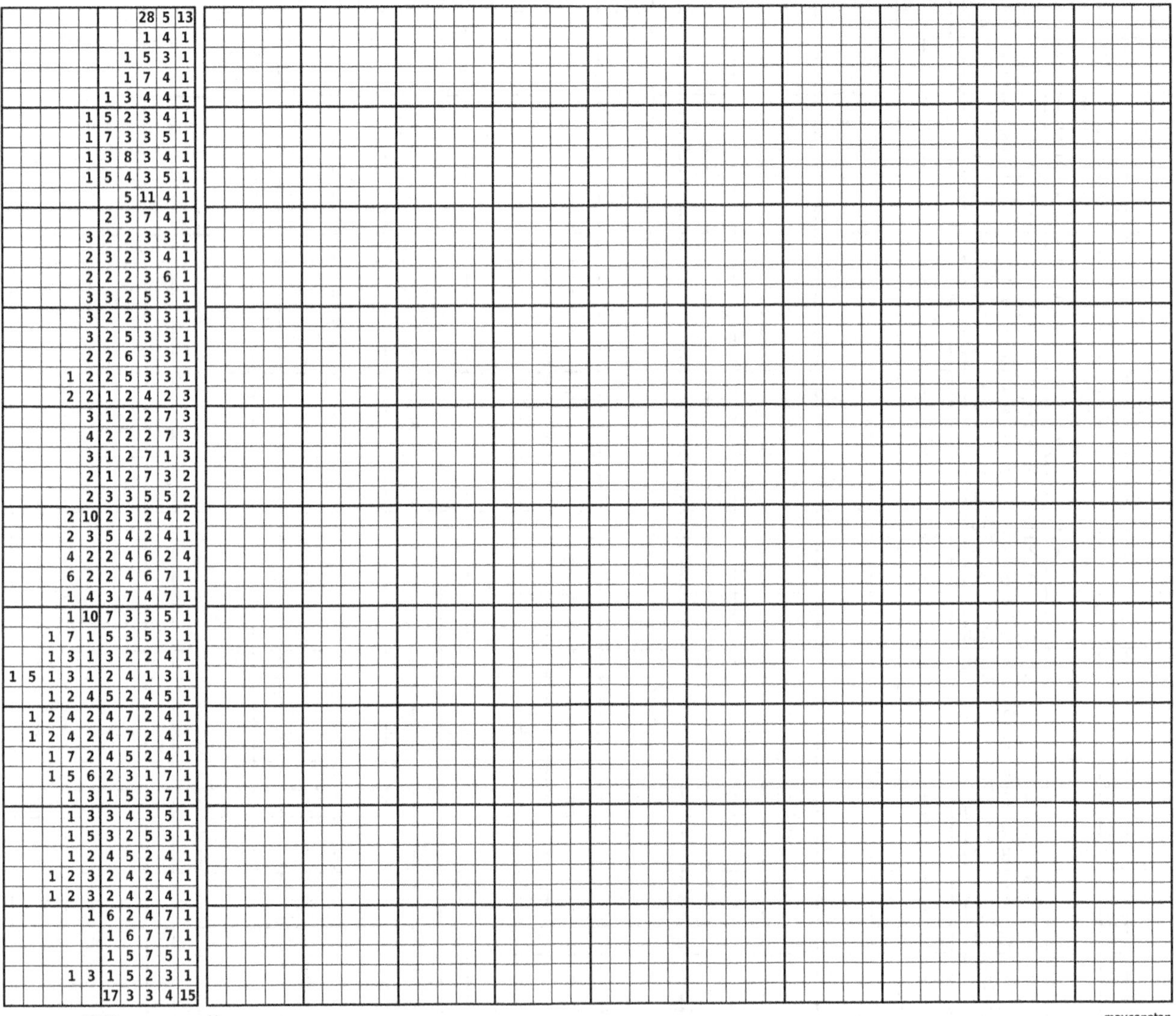

Column clues (top to bottom, columns left to right):

9 2 6 21
1 6 7 1
1 4 3 2 2 1
1 2 3 2 2 1
1 3 3 3 2 1
1 4 5 2 1
1 4 3 1
1 2 3 1
1 2 2 1
1 2 1 2 1
1 3 2 2 1
1 2 3 2 1
1 2 2 2 4 1
1 2 2 2 6 1
1 2 2 5 3 1
1 2 2 6 8 5 1
1 2 2 6 8 7 1
1 4 1 3 1 6 2 4
1 4 2 2 4 1 4 10
1 3 2 7 1 9
1 2 1 2 4 4 7 1
1 2 2 8 6 1 2
1 2 2 10 3 5 1
1 3 1 8 8 7
1 2 2 6 4 10 4
1 2 1 7 1 2 6 9
1 3 1 2 9 3 9
1 3 1 1 5 7 3 7
4 9 5 5 4 5 1
4 3 4 3 7 3 2
1 4 10 2 2 4 1 5 2
2 17 2 9 7 1
4 2 5 7 9 2 4
5 2 5 5 7 9
7 3 2 7 5 1 9
4 3 7 4 7 1
4 3 5 9 1 5 5 1
1 4 3 3 9 7 1
1 3 5 7 2 4 1
1 3 3 5 9 1
1 3 3 9 1
1 3 3 7 1
1 5 5 1
1 5 1
1 6 1
1 6 1
1 6 1
1 6 1
1 7 1
19 6 22

Row clues (left to right, rows top to bottom):

28 5 13
1 4 1
1 5 3 1
1 7 4 1
1 3 4 4 1
1 5 2 3 4 1
1 7 3 3 5 1
1 3 8 3 4 1
1 5 4 3 5 1
5 11 4 1
2 3 7 4 1
3 2 2 3 3 1
2 3 2 3 4 1
2 2 2 3 6 1
3 3 2 5 3 1
3 2 2 3 3 1
3 2 5 3 3 1
2 2 6 3 3 1
1 2 2 5 3 3 1
2 2 1 2 4 2 3
3 1 2 2 7 3
4 2 2 2 7 3
3 1 2 7 1 3
2 1 2 7 3 2
2 3 3 5 5 2
2 10 2 3 2 4 2
2 3 5 4 2 4 1
4 2 2 4 6 2 4
6 2 2 4 6 7 1
1 4 3 7 4 7 1
1 10 7 3 3 5 1
1 7 1 5 3 5 3 1
1 3 1 3 2 2 4 1
1 5 1 3 1 2 4 1 3 1
1 2 4 5 2 4 5 1
1 2 4 2 4 7 2 4 1
1 2 4 2 4 7 2 4 1
1 7 2 4 5 2 4 1
1 5 6 2 3 1 7 1
1 3 1 5 3 7 1
1 3 3 4 3 5 1
1 5 3 2 5 3 1
1 2 4 5 2 4 1
1 2 3 2 4 2 4 1
1 2 3 2 4 2 4 1
1 6 2 4 7 1
1 6 7 7 1
1 5 7 5 1
1 3 1 5 2 3 1
17 3 3 4 15

50x50

98

maycanatan

20

						1			1		1					2	4	1		1									
				1		1		1	2	1	2	3				1	2	1	3	1		1							
			1	2	1	2	1	4	1	2	1	1	2	4	3	3	3	1	2	8	6	1	1				1	1	
		1	2	1	4	3	2	3	1	1	1	3	3	4	6	2	2	2	3	1	1	4	2	1	1		1	1	
	1	3	1	2	2	1	2	2	2	4	5	1	5	1	1	1	1	1	1	1	3	5	1	3	3	1	1	1	
	1	2	1	2	4	1	2	1	1	1	1	1	1	1	2	2	3	2	3	4	1	3	1	1	1	7	1	1	
	4	2	1	2	2	1	2	1	9	1	5	2	1	1	2	2	2	2	2	1	1	1	1	2	2	2	3	1	
30	1	2	3	1	1	1	5	1	1	3	2	2	2	2	2	3	1	1	4	1	1	1	3	2	1	1	1	1	30

									30
					1	6	1	1	1
					1	2	2	5	1
				1	2	1	1	3	1
			1	2	1	1	1	2	1
		1	2	1	1	3	2	1	1
	1	1	1	1	2	2	1	2	1
	1	2	2	3	2	1	1	1	1
1	4	2	1	1	3	1	2	2	1
	1	1	3	4	2	1	1	1	1
				6	2	4	1	1	1
			1	2	2	3	1	1	1
	1	2	1	1	3	1	1	1	1
				1	3	7	2	2	1
				3	2	1	1	1	4
		2	2	1	1	1	1	2	1
		2	1	1	1	1	1	1	3
				3	2	1	3	1	1
			1	3	2	3	1	1	3
				1	3	2	2	2	1
			1	1	2	2	1	1	3
	1	1	1	1	1	2	4	1	1
		1	5	5	2	1	1	2	1
			1	1	1	1	2	5	1
			1	1	1	2	4	1	1
			1	5	2	2	1	2	1
			1	2	1	1	4	3	1
			1	2	1	2	2	5	1
			1	2	1	7	1	2	1
									30

30x30

98

xiayu

21

																2				2	1	1				2	2																						
								1	1	1				4	5	1	2	1	7	1	1	1	1	2	1	1	1	5				1	1	1				1	1	1									
						1	1	1	1	1	1	1	1	1	1	1	1	1	1	1	1	1	1	1	1	1	1	1	4	1	1	1	1	1	1	1	1	1	1	1	1	1	1						
				1	3	1	1	1	1	1	1	1	1	1	1	1	1	1	1	1	1	1	1	1	1	2	1	2	1	3	3	1	1	1	1	1	1	2	2	2	1	1	1	1	1				
		2	1	2	2	2	2	2	3	1	2	2	2	2	3	1	2	2	2	2	3	1	2	2	2	2	3	1	2	2	2	2	3	1	2	2	2	2	3	1	2	2	2	1	1	3	3	1	
4	1	3	4	2	2	2	2	2	2	2	2	2	2	2	2	2	2	2	2	2	2	2	2	2	2	2	2	2	2	2	2	2	2	2	2	2	2	2	2	2	2	2	2	5	3	2	2	1	3

							1
							1
							1
							1
							1
							6
						4	3
						2	2
						2	2
					2	2	4
					2	20	1
					2	2	4
							27
						1	2
							45
						3	2
					1	41	1
	1	6	3	3	4	4	11
					1	2	2
2	1	1	1	1	1	1	2
2	3	3	3	3	3	3	2
2	2	2	2	2	2	2	2
						2	2
							36
							34

50x25

98

twoznia

22

																									3														
												4			3							6			10	3		4											
										3	4	7	4		9		5	6	6	6	6	8	5		4	9	4	7	4	3									
	2	2							2	5	5	8	8	4	5	3	8	7	5	5	7	5	9	3	1	7	8	8	7	7	2								
	2	1	2		3	3			6	7	7	1	7	9	1	9	6	6	7	7	6	5	5	10	1	3	7	1	6	6	7								
	1	2	1	3	2	2	6	9	6	1	2	2	3	7	1	6	3	5	7	7	7	2	3	4	3	3	3	2	2	1	5	10	5				1	1	
	1	2	1	1	2	1	3	5	4	4	4	1	2	3	3	3	4	2	2	2	2	2	4	3	2	1	2	2	4	4	4	4	4	1		2	2	1	
1	5	9	19	20	5	4	3	4	1	2	2	2	1	3	3	1	1	2	2	2	2	1	1	1	2	2	1	2	1	2	1	4	4	5	1	19	11	7	
5	2	2	2	1	16	16	8	5	5	3	2	2	2	4	3	2	2	4	2	2	3	2	2	2	2	2	2	2	2	3	4	6	6	16	22	2	2	2	4
11	1	1	1	1	1	1	1	1	1	1	3	1	1	3	1	1	1	1	1	1	1	1	1	1	1	1	3	1	4	1	1	1	1	1	1	1	1	1	11

								1	7
							3	1	15
							3	2	19
								1	21
					2	2	6	7	5
						4	5	5	5
					2	1	3	4	1
					2	1	7	9	1
									27
									27
									27
									27
								4	30
					3	5	7	13	4
					1	5	5	3	1
			1	3	3	2	3	3	1
				1	5	6	6	4	7
								24	12
									34
									34
							2	25	2
							4	22	3
						5	3	3	5
				6	2	2	2	2	5
						6	2	2	6
					6	2	3	2	6
					6	4	5	4	5
					6	2	7	2	5
					6	3	7	3	5
					6	4	7	4	5
					5	4	5	4	5
							10	3	9
						5	2	2	5
						5	3	3	4
					4	3	7	3	3
							3	13	3
						5	3	3	5
					2	4	9	3	2
			2	2	2	5	3	1	2
				2	1	4	4	1	2
					1	1	12	1	1
		1	2	2	2	4	2	2	1
1	1	2	2	1	2	2	2	1	1
			1	1	2	2	2	1	1
		1	1	2	2	2	2	2	1
		1	2	2	2	2	2	1	1
				1	1	1	2	2	1
				1	2	2	2	1	1
				1	1	1	1	1	1
									40

40x50

98

stomlins123

23

50x45

98

xiayu

24

Column clues:

Column	Clues
1	18 3 4
2	1 1 2 1 1
3	1 2 2 2 2 2 1
4	1 4 5 2 2 1
5	1 1 1 4 2 2 2 2 1
6	1 3 4 4 3 2 1
7	2 2 5 4 2 2 1
8	2 2 5 2 1 2 1
9	1 2 3 7 1 2 1
10	2 4 9 1 2 1
11	4 4 7 2 2 1
12	1 5 5 4 2 1 1
13	2 2 7 1 2 1 1
14	3 2 10 1 1 1
15	8 9 1
16	6 2 7 1 1
17	1 3 1 4 2 1
18	9 2 1 1 2 1
19	7 3 1 1 2 2 2 1
20	3 3 2 3 2 1 1 1
21	2 4 3 3 3 2 1
22	9 2 3 3 2 1
23	6 4 3 3 2 2
24	4 4 3 4 3 2
25	5 4 4 2 2
26	6 4 5 2 2
27	5 4 4 2 3
28	2 4 5 2 3
29	4 5 2 3
30	2 6 1 3
31	6 2 2 1
32	4 3 2 1
33	3 2 1
34	2 1 2 1 1
35	5 1 1 1 1
36	2 2 1 1
37	2 1 4 1
38	3 5 1
39	11 1
40	2 3 4 1
41	1 3 1
42	1 2 1
43	1 2 1
44	1 1 1
45	20 1 7

Row clues:

Row	Clues
1	5 5
2	1 4 3 2 1
3	1 4 2 4 1
4	1 4 3 3 3 1
5	1 3 2 3 3 1
6	1 4 3 2 3 3 2 1
7	1 3 3 2 2 2 2 1
8	1 2 1 3 2 3 4 1
9	1 5 2 2 2 4 1
10	1 7 3 3 4 1 1
11	1 1 5 2 3 4 1 1 2 1
12	1 4 4 3 3 3 2 1 1
13	1 6 4 3 4 4 1 2 1
14	1 8 4 4 4 1 1 1
15	1 1 6 5 4 4 2 1 1 1
16	1 5 4 5 3 2 2 1
17	1 4 4 5 7 4 2 2 1
18	1 5 4 4 7 5 2 1
19	1 2 4 4 3 6 5 1
20	6 3 4 8 9 1
21	5 2 3 3 12 8
22	1 2 3 10 3 3 3
23	6 6 5 2
24	8 4 2 5 1 1
25	2 4 11 2 1 1
26	3 3 8 2 1 1
27	1 12 3 1 5 1
28	1 7 2 6 1
29	1 9 1
30	21 6 15

45x30

98

oko

25

40x40

98

muniaelfik

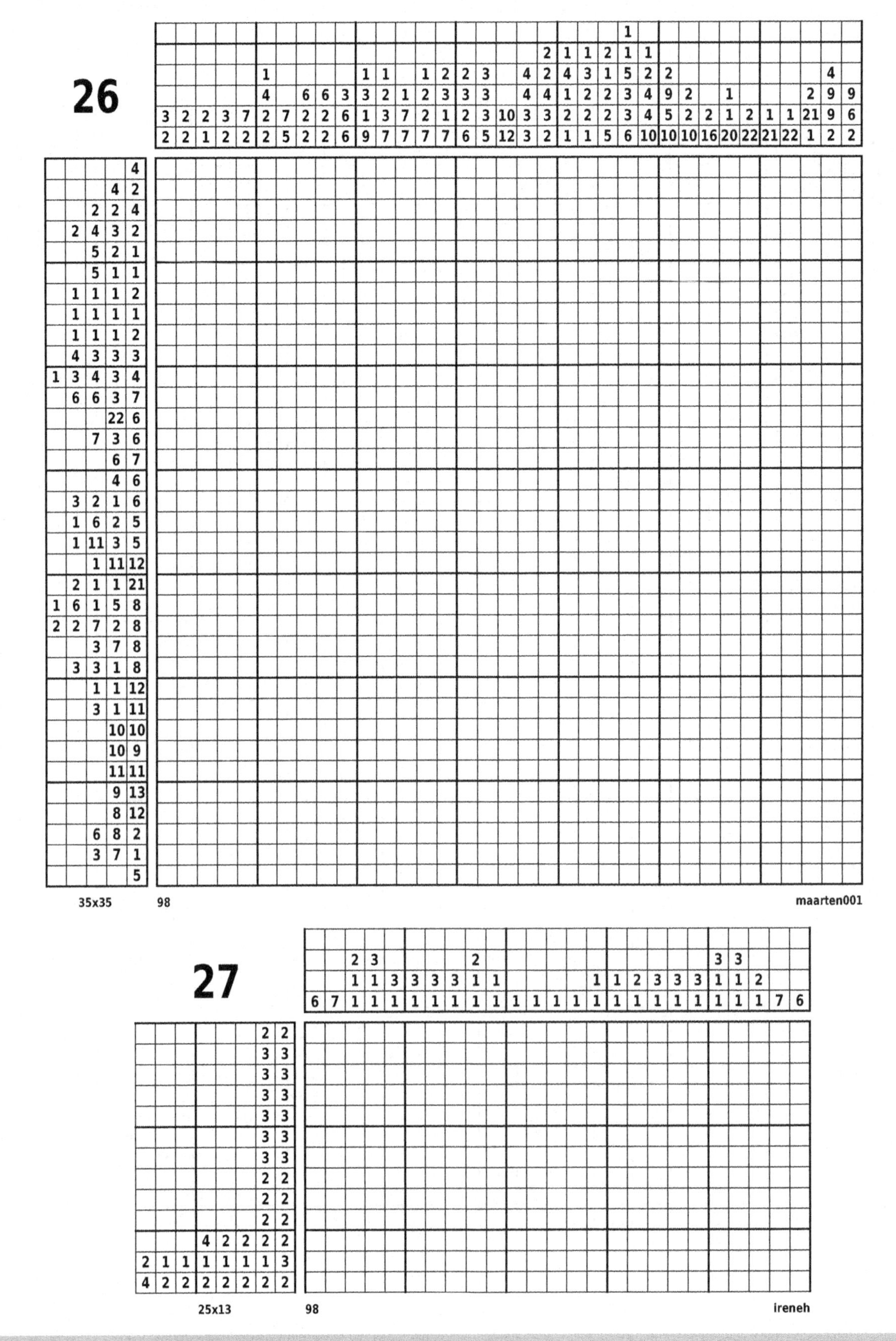
26
35x35
98
maarten001
27
25x13
98
ireneh

28

Column clues (top to bottom):

Column	Clues
1	1
2	8 1
3	3 5 2
4	2 8 5 4
5	2 1 5 7
6	2 1 5 2
7	3 1 5 2
8	4 1 5 2 1
9	5 2 2 2 3 2 1
10	7 1 2 1 2 1 2 1
11	7 1 2 2 2 1 2 2
12	8 2 2 2 3 2 1
13	9 1 2 2 1 1
14	10 1 2 2 1 1
15	10 2 4 3 1 1
16	12 1 2 3 4 1 1
17	12 2 5 2 2 1 1
18	14 1 3 3 1 1
19	14 2 3 2 1 1
20	15 1 3 2 1 1
21	16 2 3 2 4
22	17 1 6 1 1
23	17 2 5 1 1
24	19 1 1 1
25	19 2 4 3
26	20 1 2 1 1 1
27	21 2 2 1 4 1 1
28	22 1 2 2 2 2 1 1
29	21 2 2 2 1 3
30	22 1 2 2 1 1
31	22 2 2 2 1 1
32	22 1 2 3 1 1
33	21 2 7 1 1
34	21 2 5 3
35	21 1 1 1
36	19 2 4 1 1
37	18 1 1 2 1 1
38	17 2 1 3 1 1
39	16 1 2 3 1 1
40	14 2 6 3
41	13 1 4 1 1
42	12 2 3 1 1
43	11 2 5 1 1
44	10 1 2 3 1 1
45	8 2 1 2 3
46	7 1 1 2 1 1
47	5 1 2 3 1 1
48	3 2 5 1 1
49	8 1 1
50	14 2

Row clues (left to right):

Row	Clues
1	2
2	8
3	12
4	16
5	19
6	22
7	24
8	27
9	28
10	30
11	32
12	33
13	35
14	2 30
15	2 3 28
16	1 1 4 25
17	4 4 24
18	1 1 3 23
19	1 1 3 21
20	1 1 3 20
21	1 1 6 3 18
22	1 1 8 3 17
23	1 1 2 3 3 15
24	1 1 1 1 1 3 14
25	1 1 1 1 3 12
26	1 1 1 2 3 12
27	1 1 1 3 2 2 10
28	1 1 1 3 4 3 9
29	1 1 7 1 1 3 7
30	1 1 3 1 1 3 6
31	1 1 1 2 3 5
32	1 1 3 3 2 2 4
33	1 1 2 6 2 2 3 1
34	1 1 4 8 1 2 6
35	1 1 1 1 1 3 1 2 3
36	1 1 1 2 1 2 4 6 2
37	1 1 1 3 2 6 5 2
38	2 2 7 2 2 3 2
39	2 5 4 1 2 4 2
40	3 4 1 2 2 2 1
41	7 4 2 3 2 1 1
42	6 6 7 2 1 1
43	5 4 4 3 2 1
44	6 4 6 1
45	5 5 4 1
46	5 6 1
47	6 5 1
48	7 5
49	6 1
50	6

50x50

98

xxLadyJ xx

29

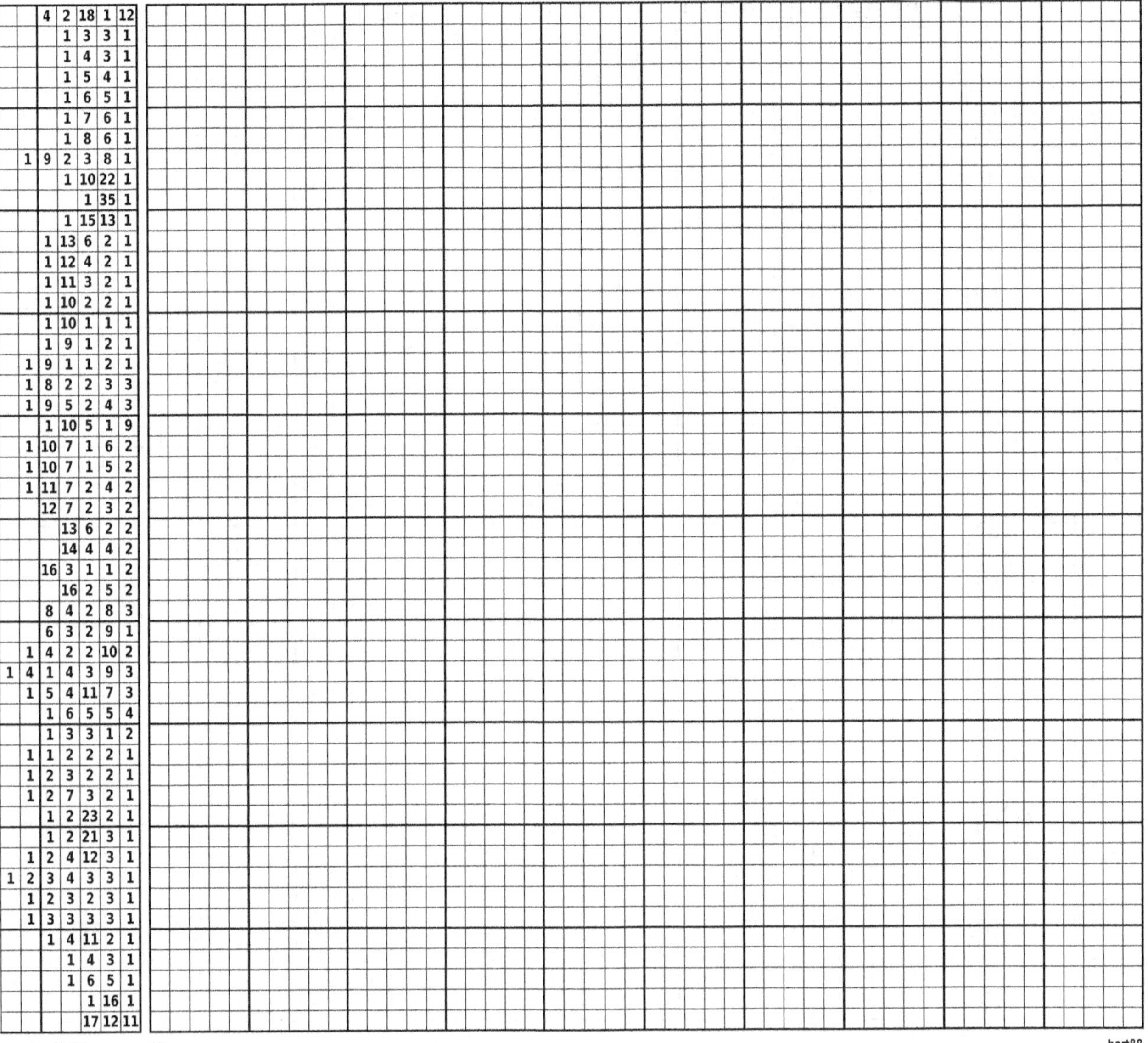

50x50

98

bart88

30

Column clues (left to right, each read top to bottom):

Column	Clues
1	50
2	19 9 3 2
3	18 4 7 4 1 3 2
4	18 2 2 6 2 2 2 1 2
5	17 1 1 4 2 1 2 1 2
6	16 2 3 3 4 1 1
7	16 2 8 3 2 1
8	15 2 5 1 1
9	9 2 2 2
10	9 2 1 2 2
11	9 4 1 2
12	8 2 5 2
13	8 2 3 1
14	8 2 1
15	8 2 1
16	8 2 2
17	2 2 1 2
18	2 2 1 1 2
19	2 3 2 1
20	2 1 5 5
21	2 1 1 4 2 2
22	2 1 6 2 2 3 2 1
23	2 1 4 6 5 2 1
24	2 2 3 3 2 3 1
25	2 1 1 2 2 1 1
26	3 1 2 2 2 1 1
27	3 2 2 2 2 1 2 1 2 1
28	3 1 4 1 1 1 1 2 2 1
29	2 2 1 3 2 2 2 1 2 1
30	1 3 2 6 2 1 2 2
31	2 1 5 2 3
32	2 2 1 2 1
33	2 2 4 1 2 2 1
34	2 2 2 1 1 4 1
35	2 2 1 1 1 1 1 1
36	2 2 1 2 1 1 1
37	1 3 2 1 1 2
38	2 1 2 1 1 2
39	3 1 2 2 3 3
40	3 1 2 1 1
41	3 1 1 4 2 1
42	2 2 1 1 3 1 1
43	2 1 8 1 6 2 1
44	2 5 1 2 3 2 6 2 2
45	2 1 1 2 2 2 3 5 1
46	9 2 2 4 3 2 3 3 1
47	9 2 4 2 2 4 2 1 2 2
48	10 5 4 4 2 3 4
49	11 15 3 2
50	49

Row clues (top to bottom, each read left to right):

Row	Clues
1	30 2 14
2	29 4 13
3	16 3 2 2 3 5
4	16 7 2 2 3 5
5	16 2 4 2 6 1 5
6	16 2 3 1 1 5
7	17 1 1 5
8	18 1 2 5
9	11 3 1 5
10	8 9 1 3
11	8 7 6 4 2
12	8 2 11 2 1
13	8 2 2 1 1 2 1
14	9 2 3 1 1 1
15	10 2 3 2 1 1
16	7 2 2 2 5 1
17	5 5 2 3 1
18	4 4 2 3 1 1
19	2 3 1 2 1 2 2
20	1 3 3 1 1 1 2
21	1 2 2 3 1 2
22	1 1 3 2 3
23	1 1 1 2 4
24	1 2 1 2 2 2 4
25	1 3 1 3 2 2 3
26	2 2 1 3 1 3 2 2
27	4 2 2 2 2 8 1 2
28	5 2 1 2 2 1 2 2
29	5 2 1 2 2 1 2 3
30	5 2 1 1 2 3 4
31	5 2 3 3 2 1 4
32	4 2 2 7 1 1 3
33	3 2 1 2 1 2 2
34	2 2 1 1 1 2 1
35	1 2 1 2 1 2 2 1
36	1 2 3 2 2 1 2 1
37	1 1 2 2 2 1 2 2
38	1 1 1 3 2 1 1 2
39	1 2 1 6 1 2 2
40	1 4 2 2 2 1
41	2 3 2 1 1 1
42	3 2 2 4 1
43	2 2 3 2 2
44	1 2 7 4 2
45	1 2 3 2 2 2 1
46	1 1 2 2 3 1 1
47	1 5 1 2 2 2 3 1 1
48	1 10 1 2 2 1 3 1 2 1
49	5 4 3 2 18 1
50	7 14

50x50

99

xxLadyJ xx

31

mustafademirbas

50x40

99

32

32x32

99

komplexyapi

33

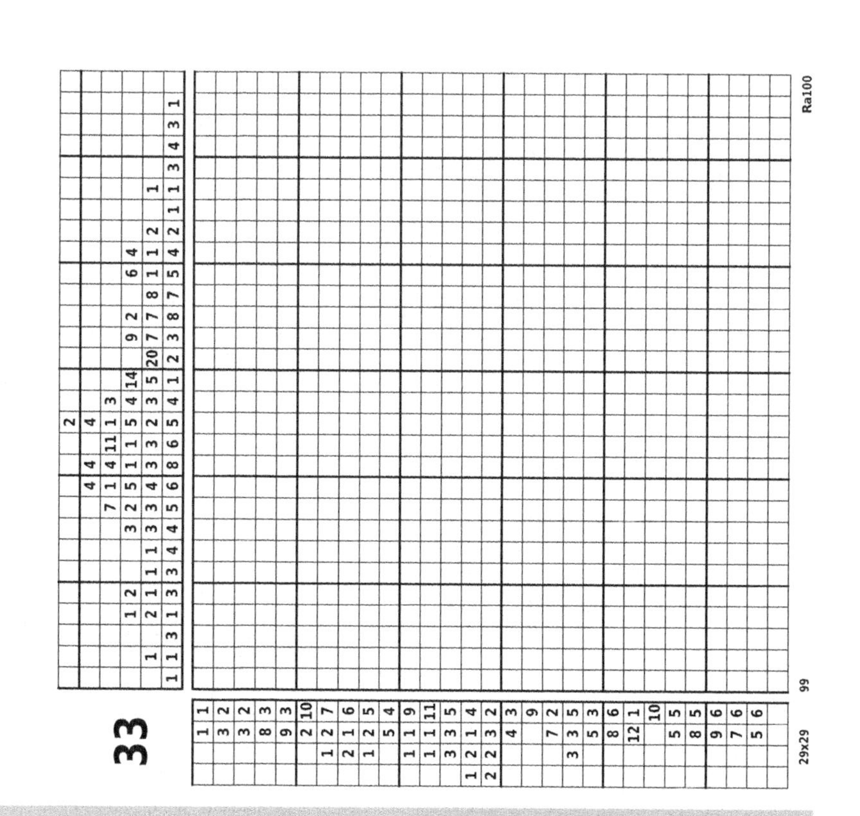

29x29

99

Ra100

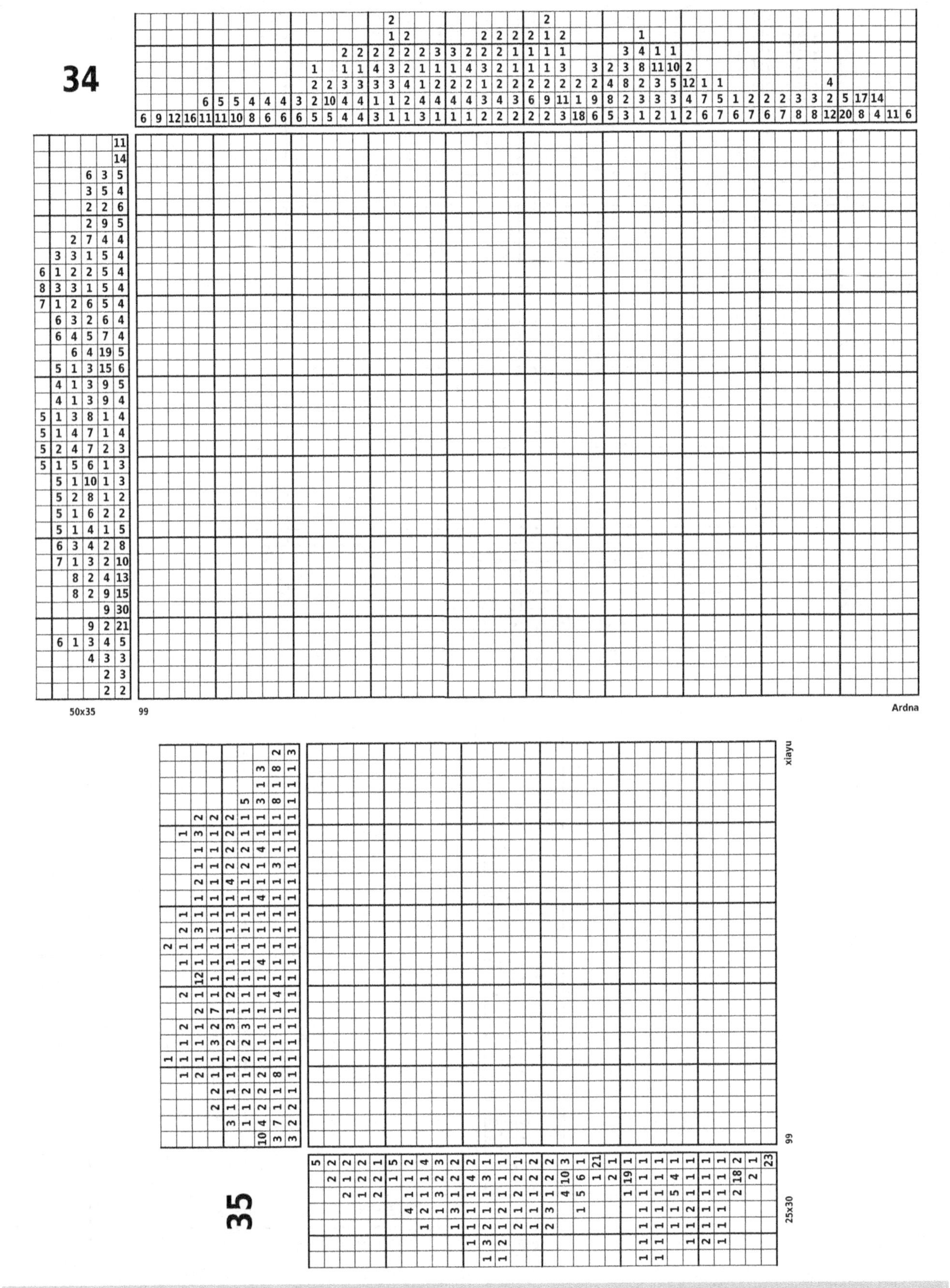
34
50x35
99
Ardna
xiayu
99
35
25x30

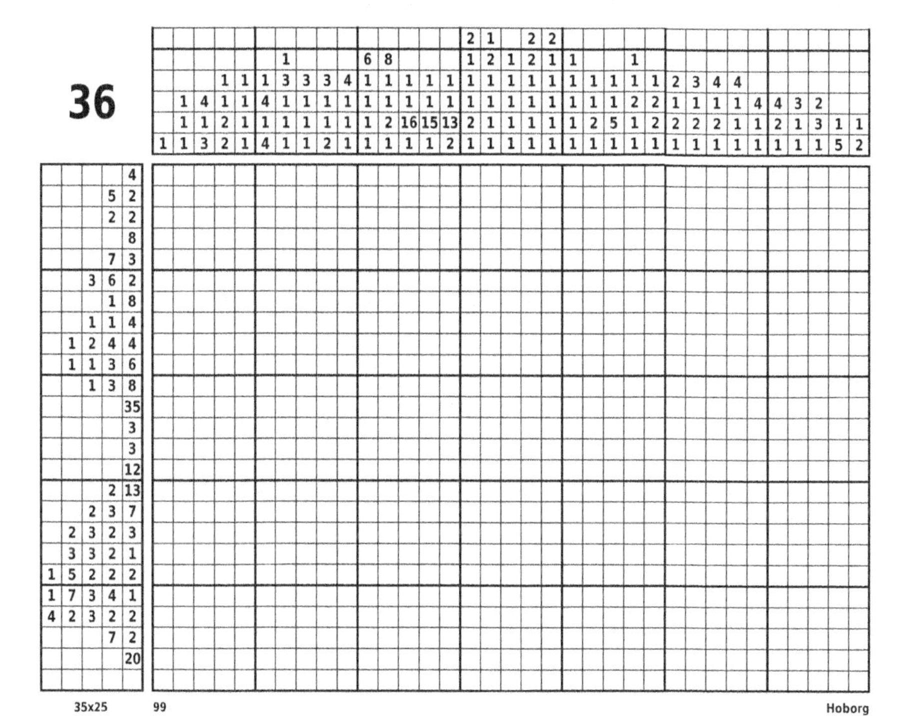

35x25 99 Hoborg

37

Column clues (left to right, top to bottom):

1. 1
2. 3
3. 4 1 2
4. 3 1 2 1
5. 2 2 1 2
6. 2 1 1 2 1
7. 2 3 1 1 2 1 3
8. 4 2 1 1 1 7
9. 1 1 1 1 9
10. 3 3 3 5
11. 5 1 6 6
12. 1 2 3 1 3 7 1
13. 8 3 2 7 2
14. 1 7 1 1 1 6 2
15. 3 3 1 2 1 5 2
16. 8 8 2
17. 1 7 9 2
18. 7 1 2 6 2
19. 1 1 4 2 2 6 2
20. 5 1 1 2 7 3
21. 5 3 2 6 2
22. 3 6 1 2 3 3
23. 2 6 2 2 7
24. 2 1 1 1 1 1 1 12
25. 2 1 1 4 9
26. 3 2 2 1 5
27. 3 3 3 1 1 2
28. 4 1 2 7 1 3
29. 1 1 2 18
30. 2 1 1 1 13
31. 1 1 2 2 1 2
32. 1 2 2 2 1 2
33. 2 1 1 2 1 2
34. 1 3 1 1 2
35. 1 1 1 1

Row clues (top to bottom):

1. 1
2. 3
3. 2 2
4. 1 1 1
5. 3 2 2
6. 2 1 2 1
7. 2 1 1 2 3
8. 1 2 3 2 1 4
9. 2 1 3 1 1 2 3
10. 1 2 2 10 2
11. 1 1 1 6 3 1
12. 1 8 8 7
13. 1 17 3
14. 1 3 7 2 4 2
15. 14 2 4 3
16. 3 2 3 3 4 2
17. 2 3 5 3
18. 2 5 2 6
19. 3 2 2 1 2 1 2
20. 9 1 3 1 2
21. 1 3 2 3
22. 2 4 1 3
23. 4 7 3
24. 6 2 2 4
25. 3 3 2 2 3
26. 2 2 2 4 2 2 1
27. 3 4 10 2 2
28. 19 3 3
29. 15 3 2 1
30. 14 8 3
31. 13 4 3
32. 10 5 5
33. 4 6 2
34. 11 4 2
35. 8 1

35x35 99 inro

38

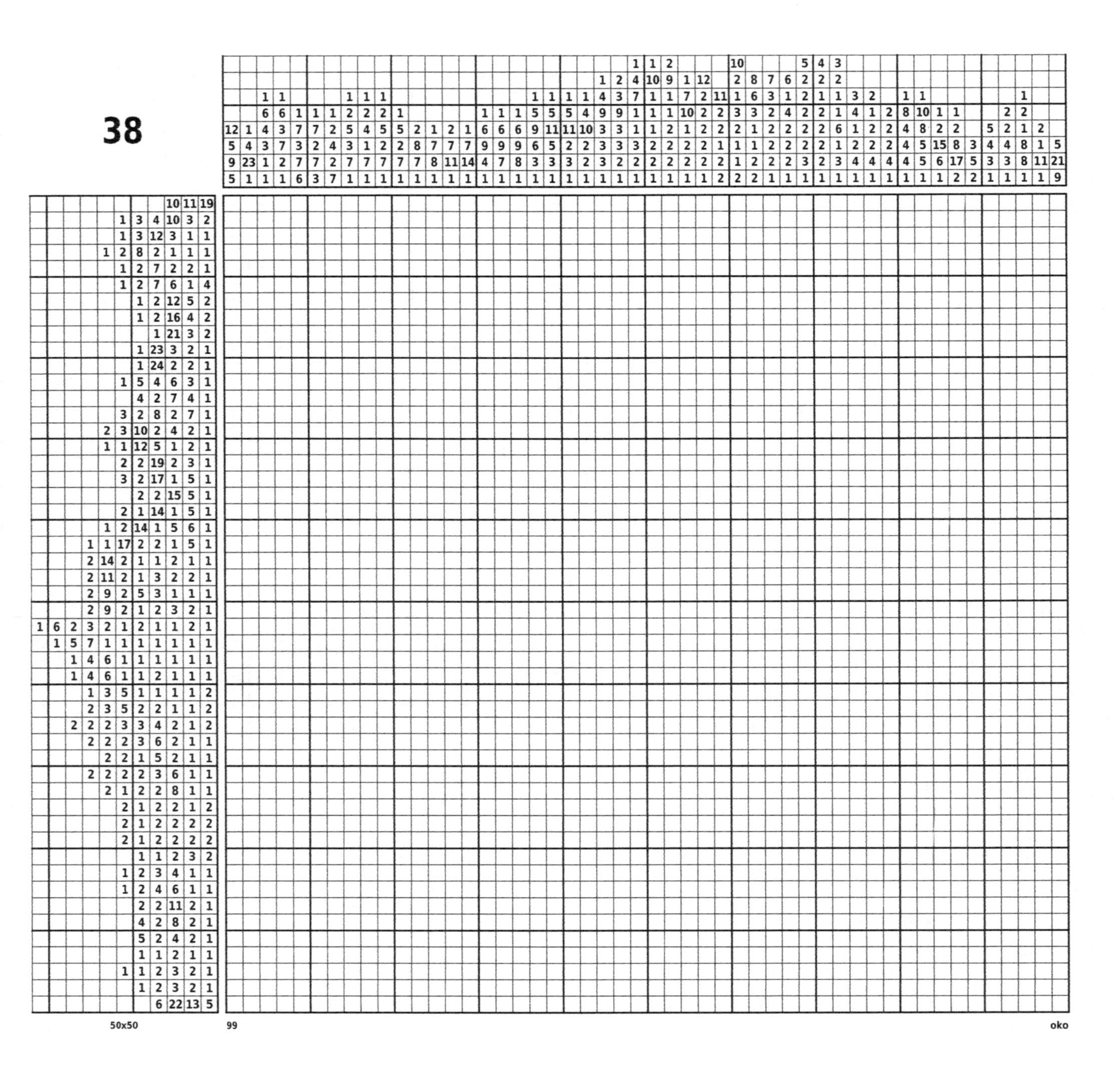

39

Column clues (columns 1–50, clues listed top to bottom):

Column	Clues
1	15
2	19
3	22
4	25
5	28
6	30
7	27 2
8	25 3
9	24 5
10	25 7
11	26 8
12	26 8
13	8 2 13 9
14	6 12 9
15	7 12 9
16	8 11 10
17	9 1 7 11
18	6 5 13
19	5 3 14
20	5 2 15
21	6 1 16
22	6 1 17
23	6 17
24	6 8 7
25	9 2 7
26	2 2 6
27	3 1 1 4
28	4 2 4 3
29	5 6 7 2
30	6 21
31	6 22
32	7 1 22
33	8 1 2 21
34	6 2 4 2 20
35	7 7 1 19
36	8 4 1 19
37	9 2 1 1 18
38	9 3 1 21
39	10 3 20
40	10 3 17
41	11 1 2 15
42	11 3 13
43	13 3 10
44	15 4 7
45	16 3 4
46	19 2
47	25
48	22
49	19
50	15

Row clues (rows 1–50, clues listed left to right):

Row	Clues
1	8 8
2	13 13
3	17 17
4	19 19
5	21 21
6	22 22
7	12 4 5 7 15
8	13 2 13 15
9	12 1 6 2 14
10	12 3 1 12
11	13 2 1 1 10
12	13 2 1 3 9
13	12 1 1 3 8
14	13 1 3 4 8
15	14 1 3 1 7
16	15 2 5 7
17	17 2 4 6
18	16 2 6
19	16 2 3 5
20	17 7 3 5
21	20 3 3 2 5
22	22 4 1 4
23	18 7 2 3
24	17 11 1 3
25	16 13 1 2
26	6 7 14 2
27	4 2 16 1
28	4 18 1
29	2 19
30	2 2 18
31	1 5 2 17
32	9 6 16
33	18 15
34	18 14
35	17 13
36	15 12
37	14 12
38	13 11
39	12 11
40	11 10
41	9 10
42	8 8
43	8 6
44	8 4
45	8 3
46	12
47	8
48	6
49	4
50	2

50x50

99

Ardna

40

Column clues (left to right, read top to bottom):

Column	Clues
1	45
2	1 1 2 2 1
3	1 2 3 3 1
4	1 2 1 2 1 1 1 1
5	1 3 1 1 2 2 1 1
6	1 1 1 1 5 1 1
7	1 1 3 1 1 1 1
8	1 2 2 1 2 2 1
9	1 3 1 1 2 1 1
10	3 2 2 1 1 2 12 1
11	2 9 1 3 1 1 5 1
12	1 2 1 1 1 1 5 3 1
13	1 2 1 3 6 1 3 2 1
14	4 4 5 2 2 1 1 1 1 3 1
15	1 4 4 5 1 1 3 1 5 3
16	1 1 2 1 1 1 1 1 1 3 1 2
17	1 2 2 5 3 1 1 4 1 1
18	1 1 2 6 1 1 3 1 1 1 3 2
19	1 3 2 1 3 1 5 1 1 1 1
20	1 1 1 1 2 4 1 1 4 1 1 5
21	1 3 2 2 1 1 2 1 3 6 1
22	1 2 1 1 1 1 2 1 1 1 2 1 1
23	1 1 3 1 4 2 1 3 4 1
24	1 2 1 1 1 2 5 2 1 1
25	1 1 3 1 1 2 1 1 1 2
26	1 1 1 1 3 1 1 2 1
27	1 2 2 2 1 3 1 3 1
28	1 2 1 1 2 1 1 2 2 1
29	1 5 3 1 3 1 1
30	1 2 1 1 6 4
31	1 3 1 3 1 1 2
32	1 3 2 3 1 3
33	1 3 1 2 2 3 1
34	1 5 1 4 1
35	45

Row clues (top to bottom, read left to right):

Row	Clues
1	35
2	1 2 1 1
3	1 2 1 1
4	1 7 3 4 1
5	5 3 2 3 1 1
6	1 3 3 1 6 1
7	1 1 1 2 2 2 1
8	1 6 1 1 2 3 1
9	3 4 3 2 1
10	4 2 2 1 5 1
11	1 2 1 1 1 2 2 1
12	1 1 1 2 3 1 1
13	1 8 1 1 1 1
14	1 2 4 2 1 1
15	1 3 1 1 2 1
16	1 2 2 2 1 1 1
17	1 3 3 1 1 1 1
18	3 1 11 1
19	8 3 1 3 1
20	1 3 7 3 2 1
21	1 4 1 4 2 1
22	1 1 4 5 2 1
23	1 2 2 1 2 5 1
24	1 1 1 5 1 1 1
25	1 4 3 1 6 2 1
26	1 1 1 1 1 2 1 2 1
27	1 1 2 4 1 4 1
28	1 1 1 1 1 1 3
29	1 1 1 6 5 2
30	1 1 1 1 1 4 5 2
31	1 6 2 1 1 5
32	1 1 1 2 1 1 1 4 2
33	1 1 1 8 1 1 1
34	1 2 1 5 1 1
35	1 1 4 2 2 1 3
36	1 1 1 1 1 2 2 1
37	1 1 1 2 3 4 1
38	1 5 1 1 6 2 1
39	1 1 1 1 1 2 1 3
40	1 2 4 1 1 1 2
41	1 3 5 1 5
42	1 1 2 6 1 3
43	1 2 2 1 3 2 1
44	1 2 1 1 1 3 1
45	35

35x45

99

xiayu

41

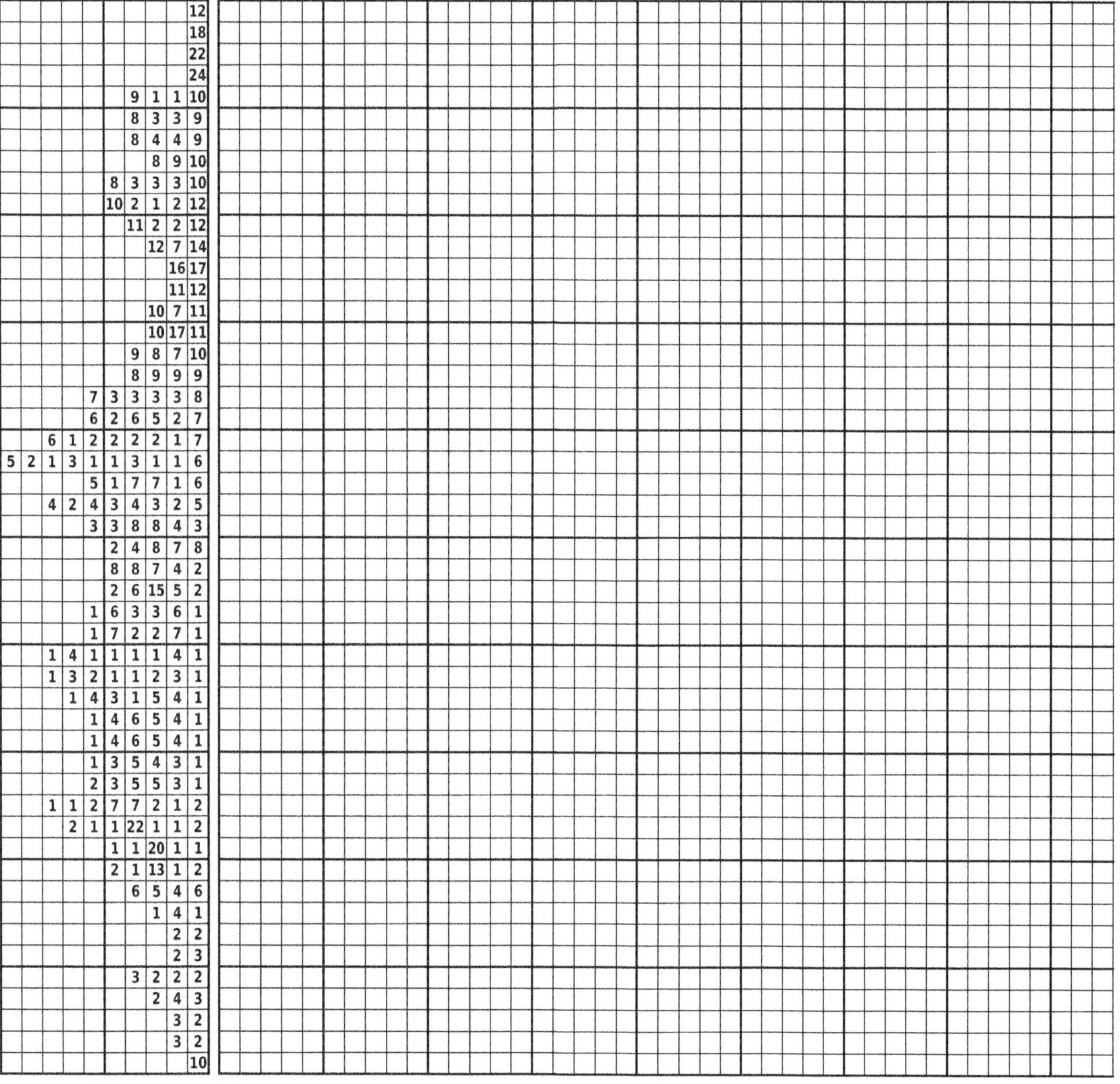

43x50

99

macro

42

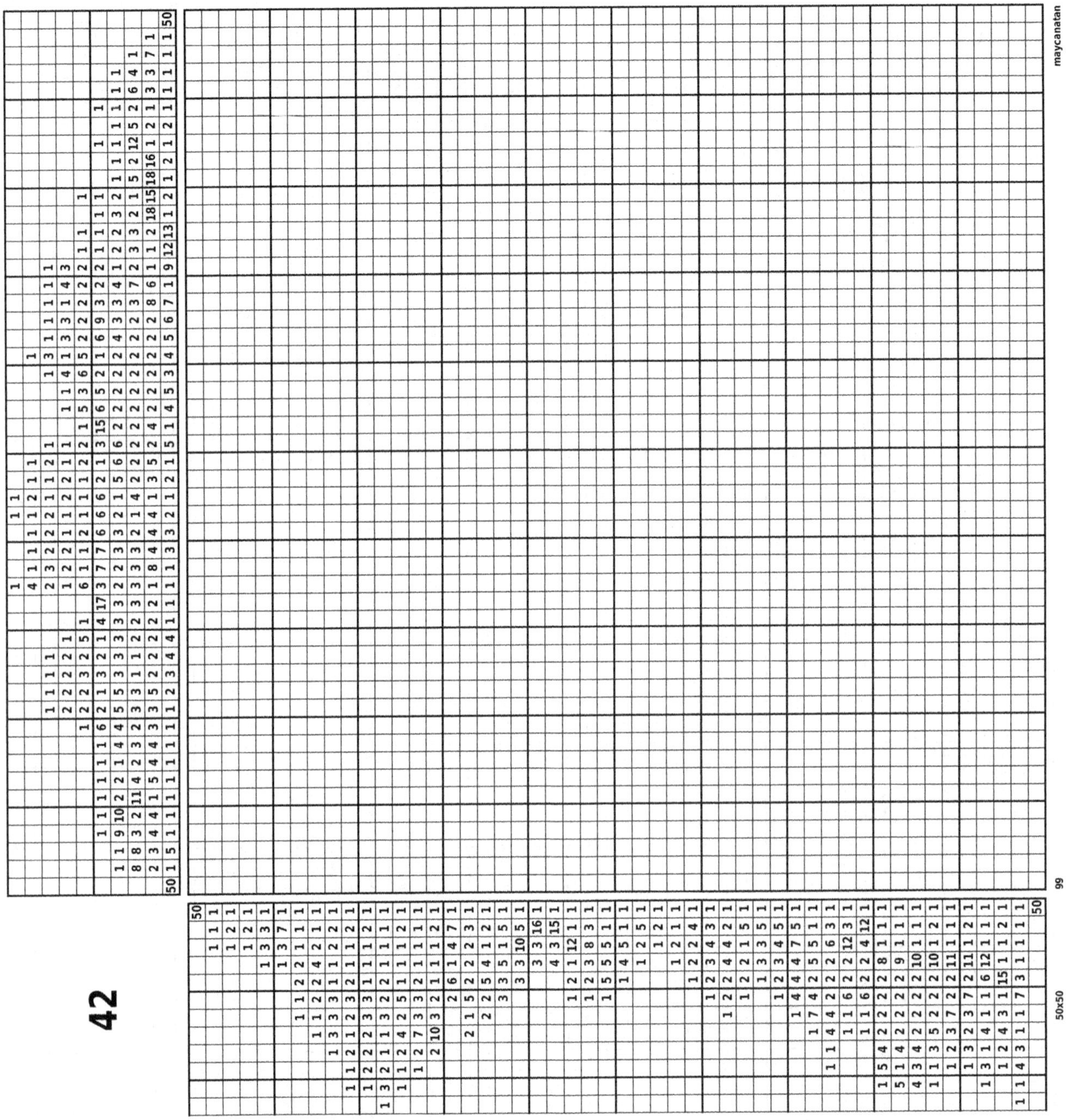

Column clues (left to right, each read top to bottom)

Column	Clues
1	50
2	1 8 2 1
3	1 8 3 5
4	1 9 3 4 1
5	1 10 2 4 1
6	1 2 11 1 1
7	1 2 4 5 1
8	1 1 2 4 1
9	1 4 3 4 1
10	1 6 4 2 3 1
11	1 2 2 2 5 3 3 1
12	1 2 2 1 5 3 5 2
13	1 2 3 3 3 1 2 3
14	1 2 2 2 3 1 2 4
15	1 5 1 3 2 2 4
16	1 4 3 2 2 1
17	17 3 3 2 1
18	1 4 2 1 6 3 2 3 1 1
19	1 3 2 1 7 2 3 8 1
20	1 2 2 1 7 3 3 4 3
21	1 2 1 2 6 3 2 4 3
22	1 1 2 1 1 6 2 1 4 2
23	1 2 1 2 1 6 1 4 1 1
24	1 1 2 1 2 5 2 3 2
25	1 2 1 2 1 6 2 5 1
26	1 1 2 3 6 2 2 5
27	1 15 2 2 4 1
28	1 5 6 2 2 2 4
29	1 3 5 2 2 2 5
30	1 4 6 2 2 2 2 3
31	1 3 1 5 1 2 2 2 4
32	1 3 2 6 4 2 2 5
33	1 3 2 9 3 2 2 6
34	1 1 2 3 3 3 8 7
35	1 4 2 2 4 7 6 1
36	1 3 2 2 1 2 1 9
37	1 1 2 3 1 12
38	1 1 2 3 2 13
39	1 3 2 18 1
40	1 1 2 1 15 2
41	1 5 18 1
42	1 2 16 2
43	1 1 12 1 1
44	1 5 2 2
45	1 1 2 1 1
46	1 6 3 1
47	1 4 3 1
48	1 7 1
49	1 1
50	50

Row clues (top to bottom)

Row	Clues
1	50
2	1 1 1
3	1 2 1
4	1 2 1
5	1 3 3 1
6	1 3 7 1
7	1 1 2 2 1 1 1
8	1 1 2 2 4 2 1 1
9	1 3 3 3 1 1 2 2 1
10	1 1 2 1 2 3 2 1 1 2 1
11	1 2 2 2 3 3 1 1 1 2 1
12	1 3 2 1 1 3 2 2 1 1 1 1
13	1 1 2 4 2 5 1 1 1 2 1
14	1 2 7 3 3 2 1 1 1 1
15	2 10 3 2 1 1 1 2 1
16	2 6 1 4 7 1
17	2 1 5 2 2 2 3 1
18	2 2 5 4 1 2 1
19	3 3 5 1 5 1
20	3 3 10 5 1
21	3 3 16 1
22	4 3 15 1
23	1 2 1 12 1 1
24	1 2 3 8 3 1
25	1 5 5 5 1 1
26	1 4 5 1 1
27	1 2 5 1
28	1 2 1
29	1 2 1 1
30	1 2 2 4 1
31	1 2 3 4 3 1
32	1 2 2 4 4 2 1
33	1 2 2 1 5 1
34	1 3 3 5 1
35	1 2 3 4 5 1
36	1 4 4 4 7 5 1
37	1 7 4 2 5 5 1 1
38	1 1 4 4 2 2 2 6 3 1
39	1 1 6 2 2 12 3 1
40	1 1 6 2 2 4 12 1
41	1 5 4 2 2 2 2 8 1 1 1
42	5 1 4 2 2 2 2 9 1 1 1
43	4 3 4 2 2 2 2 10 1 1 1
44	1 1 3 5 2 2 2 10 1 2 1
45	1 2 3 7 2 2 11 1 1 1
46	1 3 2 3 7 2 11 1 2 1
47	1 3 1 4 1 1 6 12 1 1 1
48	1 2 4 3 1 15 1 1 2 1
49	1 1 4 3 1 1 7 3 1 1 1 1
50	50

50x50

99

maycanatan

43

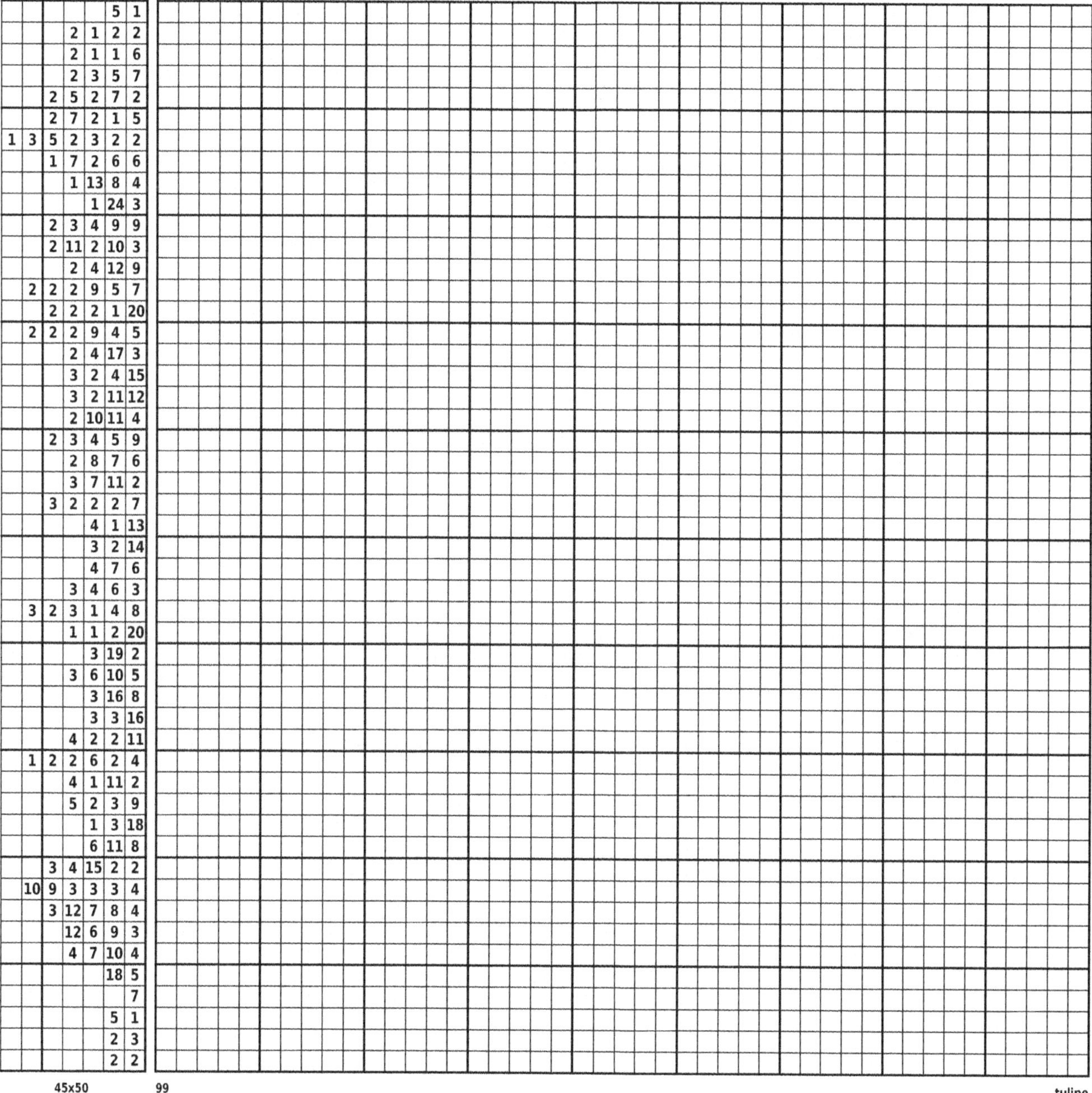

45x50

99

tulipe

44

												5																			2		4	
								3				2															2				1		1	
2		2						1	7	6	1	1	1	2	5	1		1		4		1			1		2			1	2		2	
1		1	1			1		1	2	2	2	1	2	2	2	2	5	2	1	2	4	2	1	4	1		1			2	1	4	1	2
1	2	1	2			1		2	1	5	6	1	6	1	2	2	2	2	2	2	3	2	2	2	6	2	1	1	1	1	1	1	1	1
1	4	2	1	10	10	1	8	2	1	1	1	3	3	4	2	2	3	1	3	1	3	5	6	5	1	4	1	2	2	1	1	2	1	2
1	1	4	4	10	8	1	9	1	4	4	5	2	2	2	2	2	2	2	1	7	6	1	8	10	2	4	1	10	11	1	1	9	2	2
4	5	1	1	1	2	3	4	5	1	1	1	1	1	1	1	2	5	6	10	1	1	1	1	1	1	1	1	3	3	1	2	3	1	1

						4	2
					2	1	4
				2	1	3	4
			2	1	1	1	2
			2	1	1	1	5
			2	3	1	2	1
			2	2	1	2	1
					1	4	2
					2	1	9
			3	2	1	13	3
				3	2	1	14
					2	2	14
				2	1	2	12
			3	1	2	1	2
		1	2	2	2	1	3
		1	2	1	2	2	3
		3	2	2	2	1	4
				1	2	6	4
		4	1	1	2	2	3
		1	1	5	5	2	1
			2	2	2	3	2
			5	1	6	2	2
	1	2	2	2	2	3	2
		2	1	8	2	2	2
2	1	2	2	2	1	2	2
2	1	2	3	2	1	2	1
		4	10	2	1	2	1
2	1	10	2	1	3	1	1
	3	1	3	2	3	3	2
			5	3	3	2	1
				4	12	2	1
			4	13	3	1	1
			3	4	3	6	4
			2	4	4	3	1
					26	2	2

35x35

100

xiayu

45

50x30

100

elimaor

Column clues (top to bottom):

Column	Clues
1	2
2	4 2
3	3 1 2
4	2 2 1 2
5	8 3 2
6	3 2 3 2
7	3 2 2 3 2
8	1 1 2 2 3 2
9	1 2 3 3 2
10	1 8 2 2
11	1 1 2 1 2
12	1 2 2 4 2
13	1 4 6 1
14	3 2 2 3 1
15	2 1 1 2 3 2 1
16	2 1 1 2 3 2 1
17	5 1 2 3 2 1
18	2 2 1 3 3 1
19	3 3 3 7 1
20	3 8 1 2 2
21	13 5 2
22	1 2 1 2 1 1 2
23	3 1 2 1 1 2
24	3 1 2 1 1 2
25	3 1 2 1 1 2
26	14 2 1 3
27	2 2 1 2 1 3
28	3 5 1 3 3
29	2 2 12 3
30	1 2 4 2 3
31	2 2 1 1 1 3
32	1 4 1 1 1 4
33	3 8 1 1 4
34	1 2 5 1 1 4
35	1 2 3 6 4
36	1 2 3 2 1 5
37	1 5 1 1 5
38	1 11 5 5
39	1 3 1 1 2 2 4
40	1 3 2 1 1 3 1 4
41	1 3 2 1 1 3 1 4
42	2 2 2 2 2 2 4
43	1 2 2 3 6 4
44	2 2 2 2 1 2 1 1 5
45	19 2 5
46	5 2 4 5
47	2 2 5
48	9 5
49	7 5
50	5

Row clues (left to right):

Row	Clues
1	10
2	3 3
3	3 2 2
4	2 2 2
5	2 2 4
6	4 2 6
7	8 2 6 2
8	6 4 2 5 3
9	13 2 5 1 2
10	2 1 2 3 1 5 2 2
11	1 1 1 1 1 5 1 3 2
12	13 1 1 1 4 1 2 1 2
13	2 9 2 2 1 2 1 2
14	5 3 1 2 2 1 2 2 2
15	2 2 2 1 2 1 3 3 2
16	1 3 2 1 2 1 1 2 1 2
17	2 2 1 2 3 1 2 2 13
18	3 2 1 14 3 4 3
19	1 1 1 2 11 1 7
20	4 2 2 4 1 1 3 1 2 5
21	1 9 6 7 1 1 1 2 1 2
22	1 5 2 4 4 1 1 2 2 1
23	1 2 3 1 1 11 2 1 2
24	4 2 3 1 10 2 4
25	9 3 3 4
26	6 3 3 3 7
27	4 7 19
28	5 25
29	12 31
30	50

46

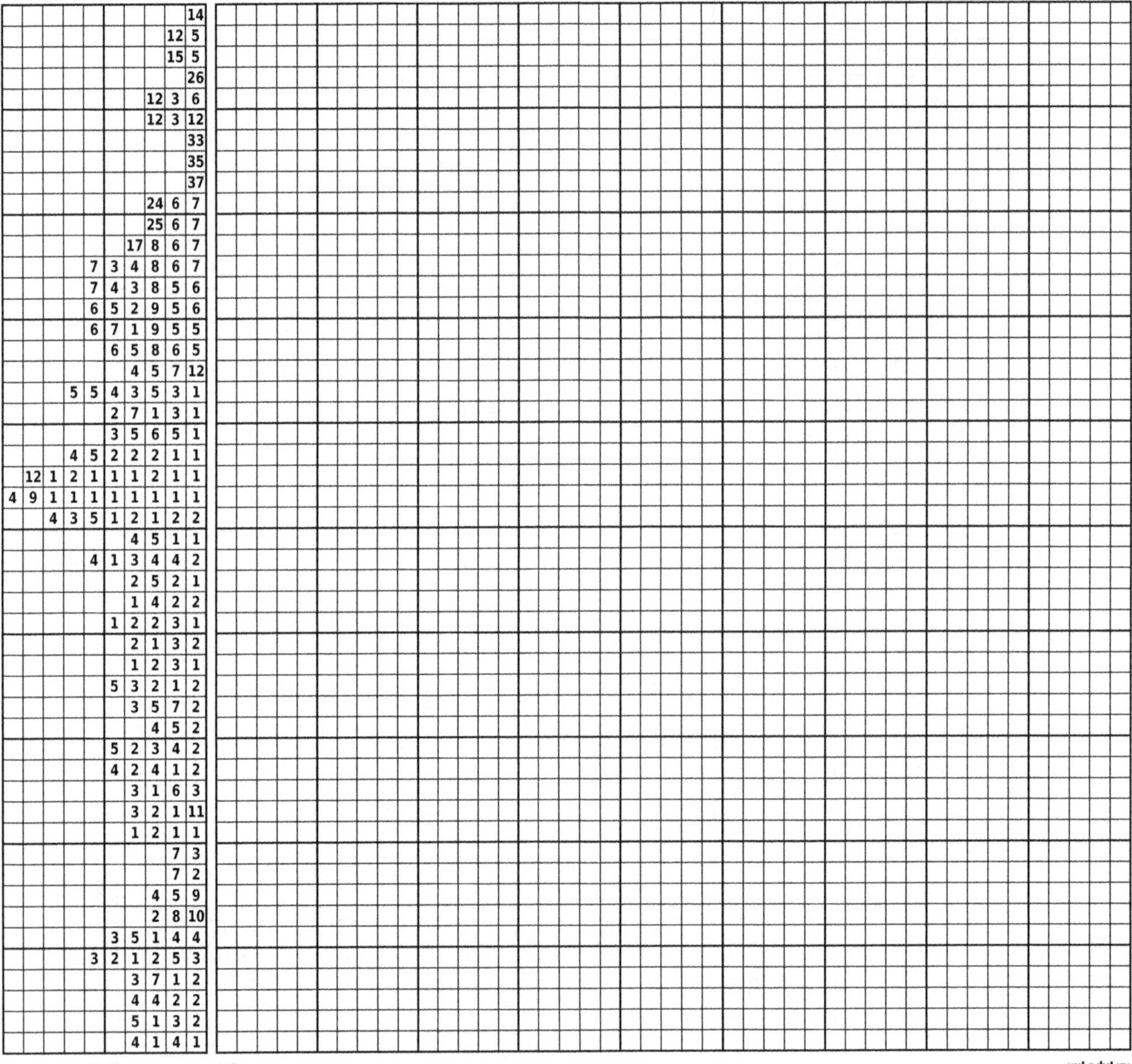

47

47x28

100

twoznia

48

Column clues (top to bottom):

Column	Clues
1	
2	2
3	25 4
4	2 1 5
5	3 17 6
6	4 19 4
7	5 4 3 2
8	6 3 2 1 2
9	5 1 2 2 4
10	2 4 2 2 2 3
11	3 3 1 1 2 2 2
12	2 2 1 2 2 2 1
13	2 1 1 1 2 2 1
14	2 2 1 2 2 2 1
15	2 1 1 1 2 2 1
16	2 2 1 2 2 2 1
17	2 1 1 1 2 2 1
18	4 1 2 2 2 1
19	2 3 1 2 2 1
20	7 2 2 1
21	5 1 2 2 1
22	7 2 2 1
23	1 3 1 2 2 1
24	7 2 2 1
25	5 1 2 2 1
26	7 3 2 1
27	2 3 1 4 3 1
28	4 1 2 19 1
29	2 1 1 1 1 17 1
30	2 2 1 2 1
31	2 1 1 1 1 3 3 7 1
32	2 2 1 2 2 2 1 1 1 1 1
33	2 1 1 1 1 1 1 1 1 1 5 1 1
34	2 2 1 2 2 2 1 1 1 1 1 1 1
35	3 1 1 1 1 3 1 1 1 7 1 2
36	2 2 1 2 1 1 1 1 4
37	1 1 1 1 3 9 6
38	2 1 2 1 5
39	1 1 27 4
40	2 7 1 1 1 1 1 1 1 1 3 2
41	1 7 1 1 1 1 1 1 1 1 3
42	8 1 1 1 1 1 1 1 1 3
43	7 1 1 1 1 1 1 1 1 6
44	6 1 1 1 1 1 1 1 1 9
45	25 5
46	4
47	2
48	

Row clues (left to right):

Row	Clues
1	2 2
2	2 2
3	2 2
4	2 2
5	2 2
6	2 2
7	2 2
8	2 2
9	2 2
10	2 2
11	37
12	5 1 1 1 3 3 1 1 1 1 1 1 1 4
13	5 1 1 1 9 1 1 1 1 1 1 5
14	5 1 1 1 11 1 1 1 1 1 1 6
15	5 1 1 1 1 9 1 1 1 1 1 7
16	5 1 1 1 1 1 1 1 1 1 1 1 1 1 1 8
17	41 1
18	1 4 2
19	1 21 3 3 1 1
20	1 23 2 2 2 1 2
21	1 4 4 1 1 1 1 1 1 1
22	1 3 3 2 2 2 1 2
23	1 2 2 3 1 1 1 1
24	1 2 2 2 1 2
25	1 2 2 7 1 1 1 1
26	1 2 2 1 1 2 1 2
27	1 2 2 7 1 1 1 1
28	1 2 2 2 1 2
29	1 2 2 2 1 1 1 1
30	1 2 2 2 1 2 1 2
31	1 2 2 2 1 1 1 1 1 1
32	1 2 2 1 3 1 2 1 2
33	1 2 2 1 1 1 1 1 1 1 1
34	1 2 2 1 1 1 1 1 2 1 2
35	1 2 2 1 1 1 1 1 1 1 3
36	1 2 2 1 3 1 2 3
37	1 3 3 2 1 1 1 4
38	1 23 2 1 6
39	1 21 2 3 3
40	1 2 3
41	37 3
42	3 3 3 3
43	3 3 3 3
44	3 2 3 2
45	3 3
46	3 3
47	3 3
48	2 2

48x48

100

tworznia

49

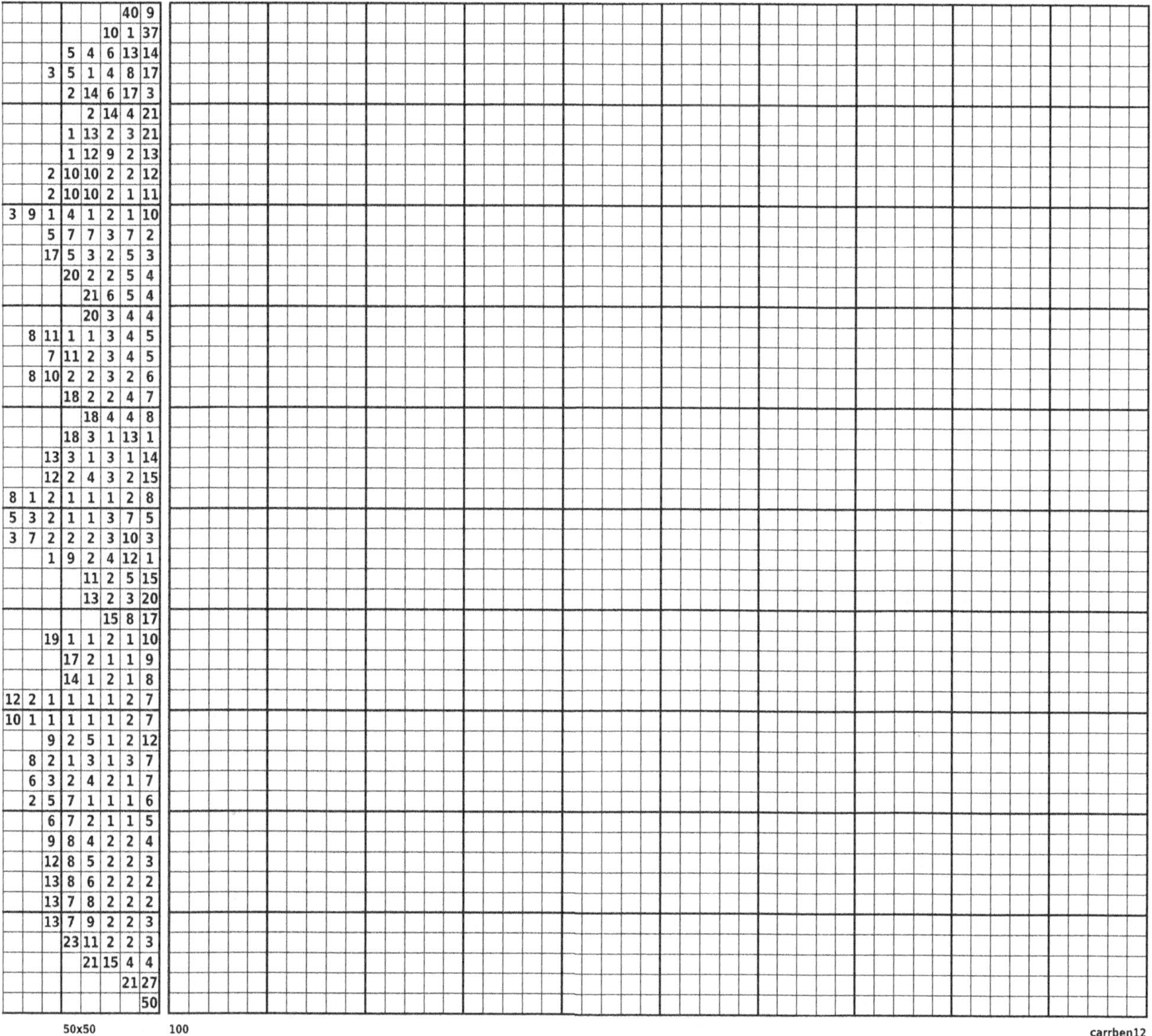

50

Column	Clues
1	3
2	4 6
3	4 3 3
4	4 8 3
5	5 2 2 3
6	6 2 6 4
7	6 2 2 2 4
8	6 2 6 2 3
9	3 3 2 9 1 4
10	7 2 1 8 1 3
11	8 2 11 2 4 1
12	11 4 8 4 1
13	7 7 4 3 4 2
14	5 3 3 3 6 4 2
15	12 3 2 3 19
16	9 4 2 3 19
17	7 4 2 2 4 2
18	4 2 2 4 1
19	4 2 6
20	4 3 5
21	4 2 3 1
22	4 6
23	5 6
24	5 6
25	5 7
26	6 6
27	6
28	7
29	8
30	7

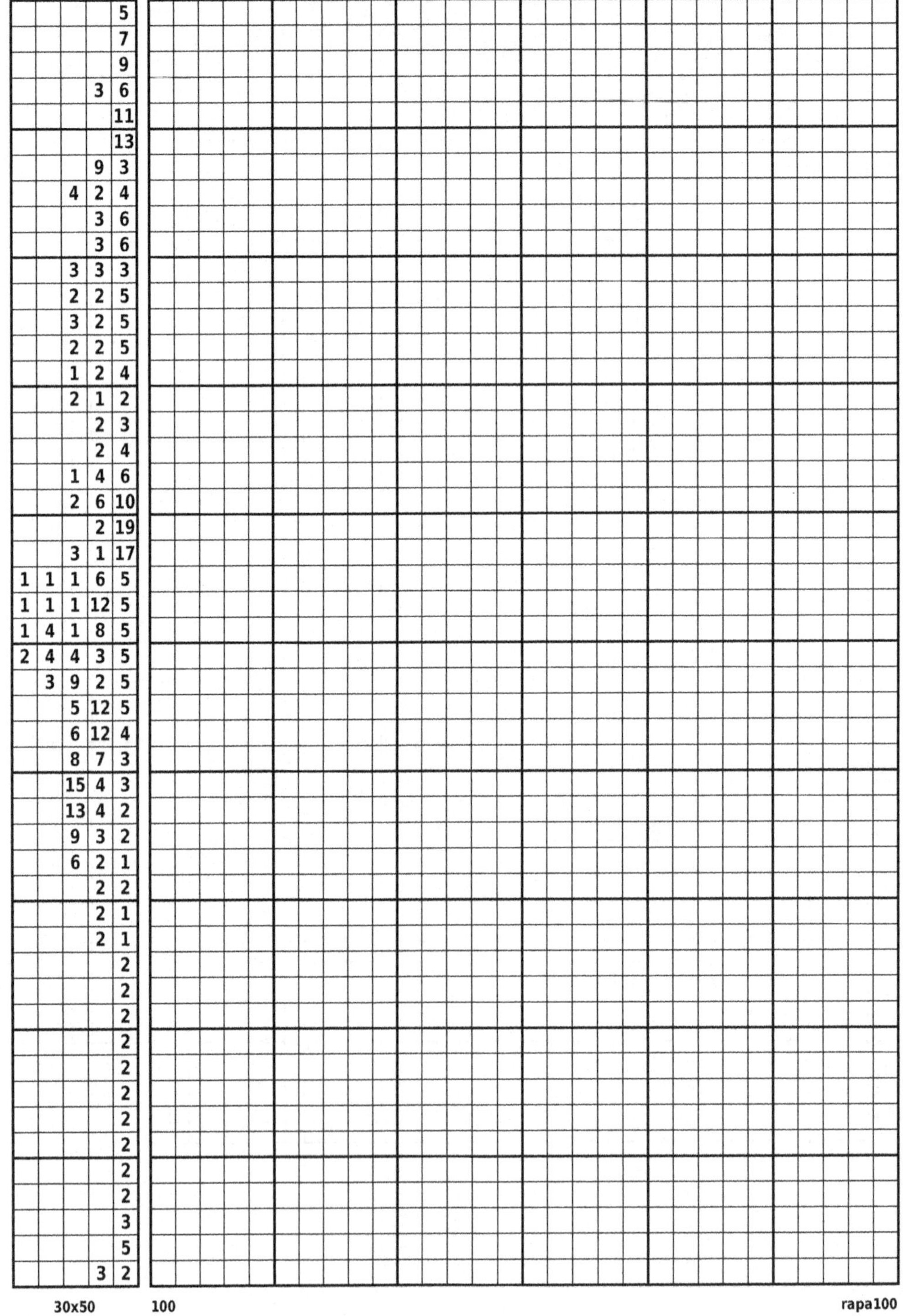

Row	Clues
1	5
2	7
3	9
4	3 6
5	11
6	13
7	9 3
8	4 2 4
9	3 6
10	3 6
11	3 3 3
12	2 2 5
13	3 2 5
14	2 2 5
15	1 2 4
16	2 1 2
17	2 3
18	2 4
19	1 4 6
20	2 6 10
21	2 19
22	3 1 17
23	1 1 1 6 5
24	1 1 1 12 5
25	1 4 1 8 5
26	2 4 4 3 5
27	3 9 2 5
28	5 12 5
29	6 12 4
30	8 7 3
31	15 4 3
32	13 4 2
33	9 3 2
34	6 2 1
35	2 2
36	2 1
37	2 1
38	2
39	2
40	2
41	2
42	2
43	2
44	2
45	2
46	2
47	2
48	3
49	5
50	3 2

30x50 100 rapa100

51

Column clues (top to bottom, columns 1–50):

Column	Clues
1	4
2	6
3	2 2
4	1 2
5	2 3
6	10
7	7 4
8	5 3
9	1 3
10	1 2
11	2 2
12	1 2
13	1 1 1 1 2
14	2 1 1 2 2
15	1 2 1 1 3 2
16	2 6 3 2
17	3 10 2
18	10 2 3
19	4 7 1 1 1 3
20	9 1 4 4
21	8 1 3 2 2 2
22	12 1 3 2 2
23	7 1 3 6
24	9 1 3 6
25	12 1 6
26	10 6
27	9 6
28	14 2
29	9 2
30	15 2 2
31	12 2 2
32	11 6
33	16 6
34	13 1 7
35	15 7
36	20 8
37	12 7
38	12 10
39	15 11
40	2 12 12
41	1 28
42	2 24 2
43	2 24 1
44	2 2 20 2 1
45	3 19 1 1
46	2 2 15 1 2 1
47	2 2 10 1 2 1
48	2 2 2 1 2 1 1 2 1
49	2 1 2 1 1 2 1
50	2 1 1 1 1

Row clues (left to right, rows 1–35):

Row	Clues
1	1 1 1 1 1 1 1
2	2 1 1 1 1 1 1 1 2 1
3	2 1 1 1 1 1 1 1 2 2 1
4	2 1 2 1 3 1 2 2 1 1 2 2
5	2 2 3 5 5 2 1 2 1 1
6	2 22 1 2 2
7	2 29 2
8	29 2 1
9	30 2
10	36
11	32 1
12	30 3
13	2 23
14	2 20 2
15	1 4 20
16	6 2 2 15
17	3 4 2 2 2 13
18	2 3 2 1 2 2 9
19	2 3 2 1 2 2 7
20	3 4 4 3 3 10
21	10 8
22	4 11
23	1 9 2
24	2 2 11
25	2 2 9 2
26	2 3 11
27	3 3 8 2
28	3 3 11
29	4 3 10 2
30	34
31	29 3
32	5 5 5
33	5 5 3
34	7 7
35	7 7

50x35

100

anawa74

52

50x44

Column clues (left to right, top to bottom):

Column	Clues
1	16
2	5 2 7
3	5 4 2 6
4	5 1 1 1 6
5	7 1 1 2 5
6	7 4 1 5
7	3 3 3 1 4
8	2 2 3 2 4
9	6 2 4 1 4
10	8 3 4 1 3
11	5 2 1 2 4 1 3
12	5 2 1 2 4 1 3
13	3 3 2 2 3 2 3
14	2 1 4 2 3 4 1 3
15	2 1 5 2 3 4 1 3
16	1 1 8 2 2 3 1 3
17	2 11 2 2 3 1 2
18	1 1 7 2 2 3 1 2
19	2 1 8 2 2 1 3 2
20	1 2 8 2 2 1 1 1 2
21	1 10 2 3 4 1 2
22	1 11 1 3 2 2 2
23	2 11 1 3 2 1 2
24	1 11 2 3 2 2 2
25	1 10 2 3 2 2 3
26	1 11 1 3 2 2 3
27	2 11 4 2 1 4
28	1 12 2 2 5 5
29	1 13 2 3 1 7
30	1 9 3 2 3 5
31	2 5 2 4 5
32	2 9 2 2 7 5
33	2 10 3 7 6
34	3 10 1 7 7
35	3 9 1 4 8
36	3 7 2 11
37	4 4 1 2 16
38	4 1 3 10 17
39	5 1 3 18
40	4 3 2 25
41	4 2 1 3 19
42	5 1 3 2 3 18
43	5 1 2 6 19
44	6 1 1 4 20
45	6 1 3 21
46	8 2 24
47	13 29
48	10 30
49	44
50	44

Row clues (top to bottom):

Row	Clues
1	24
2	5 20
3	5 17
4	3 14
5	3 1 9
6	2 7
7	2 2 2 5
8	2 1 2 1 5
9	9 1 1 1 4
10	2 1 5 2 1 1 1 4
11	2 2 8 5 2 1 2
12	2 2 12 5 3 1 2
13	2 15 5 3 1 2
14	2 16 5 2 1 2
15	2 17 5 1 3
16	2 18 6 3 4
17	22 5 1 2 4
18	22 4 1 2 2 4
19	2 14 3 1 1 1 1 4
20	10 9 2 1 2 3 4
21	2 9 4 1 1 1 3 5
22	3 9 2 1 1 1 3 5
23	3 4 1 1 1 1 3 5
24	4 6 1 1 1 6
25	4 3 4 1 1 7
26	5 3 2 2 2 2 8
27	6 3 2 2 3 12
28	2 6 3 2 2 1 13
29	2 9 4 2 2 1 14
30	5 14 2 2 3 14
31	1 1 1 9 2 1 4 14
32	1 1 1 2 2 4 14
33	1 1 7 2 1 5 14
34	2 12 4 1 4 15
35	3 12 1 2 4 15
36	1 3 10 1 3 4 15
37	1 4 5 1 2 1 1 16
38	2 6 2 2 2 17
39	4 7 3 1 18
40	6 4 23
41	9 24
42	16 26
43	50
44	50

100

xxLadyJ xx

53

50x50 — 100 — tulipe

Column clues (columns 1–50, top to bottom)

Column	Clues
1	8 38
2	10 1
3	2 8 1
4	1 7 1
5	1 4 1
6	1 2 24 1
7	6 2 6 3 3 1
8	6 2 4 2 1 1
9	6 4 2 2 2 1
10	7 8 9 1 1 1
11	3 7 4 14 2 2 1
12	2 6 14 6 1 1
13	2 6 13 2 5 1
14	2 7 11 2 1
15	12 6 12 2 1
16	1 8 5 13 1 1
17	1 14 2 1
18	1 11 1 1
19	1 8 2
20	1 20 10
21	33 4 2
22	6 5 5 15 3 1
23	3 4 4 6 2 2
24	2 1 3 4 2 1 1
25	2 2 2 3 3 1 2
26	2 3 2 3 2 6 1
27	1 5 1 2 2 8 9
28	1 3 1 2 2 9 1
29	1 2 1 2 4 4 1 1
30	1 2 1 3 1 1
31	1 1 1 1 3 1 1
32	1 2 3 2 1 1
33	1 3 3 2 1 1
34	1 5 3 2 1 1 1
35	1 6 5 2 1 5
36	1 1 6 3 2 5
37	1 1 7 6 2 1
38	1 2 11 2 3 1 2
39	1 3 8 2 3 2 1
40	1 5 7 2 3 2 2
41	1 7 6 2 4 4 1
42	1 1 5 1 19
43	1 2 2 2 2
44	1 3 1 2 2 1
45	1 4 3 1 4 1
46	6 3 2 3 1
47	2 4 1 3 3 1
48	2 2 3 3 1
49	1 7 1
50	24 23

Row clues (rows 1–50, left to right)

Row	Clues
1	11 5 2 19 3
2	2 8 3 2 1
3	2 5 2 2 2 2 1
4	3 4 2 1 2 2 1
5	4 4 3 1 2 3 1
6	10 5 2 4 5 1
7	6 7 2 22 1
8	5 6 2 2 4 4 2 1
9	5 6 1 2 3 3 2 1
10	4 6 1 2 2 2 1 1
11	3 6 1 2 2 1
12	1 1 2 1 1 1
13	1 2 1 3 1 1
14	1 1 2 5 2 1
15	1 1 2 10 2 2 1
16	1 5 7 3 2 1
17	1 7 2 3 1
18	1 13 8 1
19	1 14 10 1
20	1 3 13 10 2
21	1 2 20 9 4
22	1 2 19 8 7
23	1 1 4 10 6 5
24	1 1 5 6 4 2
25	1 1 4 6 2 1
26	1 1 4 3 3 1 1 2
27	1 1 4 4 3 1 1 2 2
28	1 1 3 4 3 1 2 6 1
29	1 1 3 3 4 2 1 4 1
30	1 2 2 3 5 2 1 2 1
31	1 2 2 3 7 3 2 2 1
32	1 2 2 2 4 20 1
33	1 3 1 2 4 19 1
34	1 3 1 2 3 3 1 1
35	1 3 1 2 2 3 1 1
36	1 1 2 2 2 3 2 1
37	1 1 1 1 1 2 2 4 1
38	1 2 2 1 2 2 5 1
39	1 3 2 1 3 7 1
40	1 4 3 1 1 2 5 1 1
41	1 1 2 1 2 1 1 1 1
42	1 1 1 3 1 1 1 1 1 1
43	1 1 2 5 1 1 2 1
44	1 1 1 4 9 4 1
45	1 2 1 6 1 1 5 1
46	1 1 1 1 1 4 1 2 2 1
47	1 7 1 1 1 1 1
48	1 1 1 1 1 1
49	1 2 1 1 1 1 1 1 1 1
50	18 7 6 7 7

54

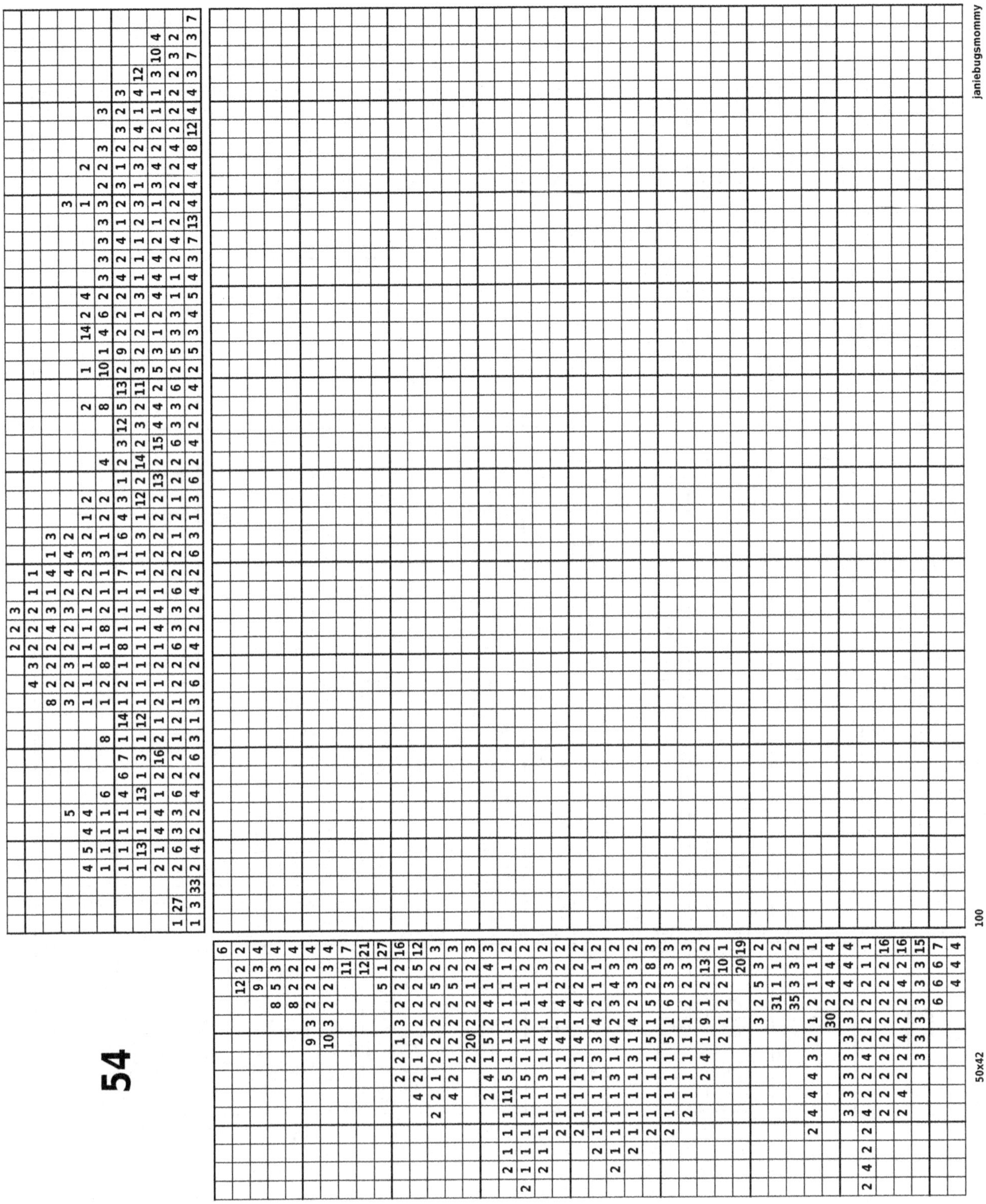

Column clues (columns 1–50, top to bottom)

Column	Clues
1	1 1
2	27 3
3	33
4	4 1 1 1 2 2 2
5	5 1 1 13 1 6 4
6	4 1 1 1 4 3 2
7	5 4 1 1 1 4 3 2
8	6 4 13 1 6 4
9	6 1 2 2 2
10	7 3 16 2 6
11	8 1 1 2 1 3
12	14 12 1 2 1
13	8 3 1 1 1 1 2 1 3
14	4 2 2 1 2 2 1 1 2 6
15	3 2 3 1 8 1 1 2 2 2
16	2 2 2 2 1 1 8 1 1 6 4
17	2 2 4 2 1 8 1 1 4 3 2
18	3 2 3 3 1 2 1 1 4 3 2
19	1 1 2 2 1 1 1 1 6 4
20	1 4 4 2 1 7 1 2 2 2
21	1 4 3 3 1 1 2 2 6
22	3 2 2 1 6 3 2 1 3
23	1 2 4 1 2 2 1
24	2 2 3 12 2 1 3
25	1 2 13 2 6
26	4 2 14 2 2 2
27	3 2 15 6 4
28	12 3 4 3 2
29	2 8 5 2 4 3 2
30	13 11 2 6 4
31	1 10 2 3 5 2 2
32	1 9 2 3 5 5
33	14 4 2 2 1 3 3
34	2 6 2 1 2 3 4
35	4 2 2 3 4 1 5
36	3 4 1 4 1 4
37	3 2 1 4 2 3
38	3 4 1 2 4 7
39	3 1 2 1 2 13
40	3 1 3 2 3 1 2 4
41	2 3 1 3 2 4
42	2 2 1 3 4 2 4
43	3 2 2 2 4 8
44	3 4 2 2 12
45	3 2 1 1 2 4
46	3 4 1 2 4
47	12 3 2 3
48	10 3 7
49	4 2 3
50	7

Row clues (rows 1–42, left to right)

Row	Clues
1	6
2	12 2 2
3	9 3 4
4	8 5 3 4
5	8 2 2 4
6	9 3 2 2 2 4
7	10 3 2 2 3 4
8	11 7
9	12 21
10	5 1 27
11	2 2 1 3 2 2 2 16
12	4 2 1 2 2 2 2 5 12
13	2 2 1 2 2 2 2 5 2 3
14	4 2 1 2 2 2 5 2 3
15	2 20 2 2 1 2 3
16	2 4 1 5 2 4 1 4 3
17	2 1 1 1 11 5 1 1 1 1 1 1 2
18	2 1 1 1 1 1 5 1 1 2 1 1 2 2
19	2 1 1 1 1 3 1 4 1 4 1 3 2
20	2 1 1 1 1 4 1 4 2 2 2
21	2 1 1 1 1 4 1 4 2 2 2
22	2 1 1 1 1 3 3 4 2 1 1 2
23	2 1 1 1 1 3 1 4 2 3 4 3 2
24	2 1 1 1 1 3 1 4 2 3 3 2
25	2 1 1 1 1 5 1 5 2 8 3
26	2 1 1 1 1 5 1 6 3 3 3
27	2 1 1 1 1 1 2 2 3 3
28	2 4 1 9 1 2 13 2
29	2 1 2 2 10 1
30	20 19
31	3 2 5 3 2
32	31 1 1 2
33	35 3 3 2
34	2 4 4 4 3 2 1 2 1 1 1
35	30 2 4 4 4
36	3 3 3 3 3 3 2 4 4 4
37	2 4 2 2 4 2 2 4 2 2 2 2 1 1
38	2 2 2 2 2 2 2 2 2 16
39	2 4 2 2 4 2 2 4 2 16
40	3 3 3 3 3 3 3 15
41	6 6 6 7
42	4 4 4

50x42

100

janiebugsmommy

55

Column clues (top to bottom):

Column	Clues
1	6
2	2 10
3	2 1 2 1
4	2 3 2 1
5	4 6 4 1 1 1
6	2 5 5 1 1 2
7	1 3 3 1 3 2
8	1 2 4 3 2 4 2
9	1 2 6 4 2 2 4 2
10	1 1 2 1 4 2 2 3 2
11	2 1 2 1 2 4 2 3 3 2
12	1 1 2 2 1 4 2 4 2 3
13	2 1 1 1 1 4 2 1 1 1 4
14	2 2 2 4 4 1 1 1 2 5
15	1 3 3 2 3 1 2 1 2 5
16	1 3 6 1 3 1 2 1 3 5
17	1 2 1 2 4 1 3 4
18	1 2 5 4 2 3 3
19	1 10 5 2 5
20	1 1 3 5 2 4
21	1 1 6 2 3
22	2 15 3 4
23	2 1 3 6
24	4 4 5 1 4
25	13 6 1 4
26	1 21 1 4
27	2 9 1 4 1 3
28	2 3 3 1 3
29	2 2 3 1 3
30	2 2 2 1 2
31	4 2 4 4 1
32	6 5 3 1
33	7 3 1
34	7 3
35	10 8 3
36	7 7 4
37	16 3
38	13 12
39	6 5 2 2
40	1

Row clues (left to right):

Row	Clues
1	10
2	2 3
3	3 2 1
4	6 5 1
5	2 6 2
6	1 5 2 1
7	1 2 3 1 1 3
8	2 1 6 1 1 1 2
9	2 1 4 2 1 1 2 2
10	2 1 1 2 2 1 2 1 2 2
11	2 2 1 2 1 1 2 1 3 2
12	1 2 1 3 1 3 1 4 1
13	1 2 1 1 2 2 1 4 2 1
14	1 2 5 5 6 2 1 2
15	1 2 1 1 4 1 1 2
16	4 11 1 3 1 1 2
17	1 1 10 1 3 1 1 2
18	1 3 9 2 2 2 2 2
19	1 2 7 2 2 2 4
20	1 2 3 1 2 4
21	1 3 4 1 2 5
22	1 3 5 1 1 5
23	1 15 4 1 5
24	1 6 4 3 2 4
25	1 5 3 3 3
26	1 8 5 5 2
27	2 5 8 6
28	6 2 11 6
29	8 4 6
30	3 6
31	2 6
32	5 2 6
33	8 1 5
34	15 2 2
35	6 15 1
36	5 6 2 3 1
37	5 6 5 4 1
38	7 7 5
39	12 8 5
40	14 11 5

40x40

100

Id5

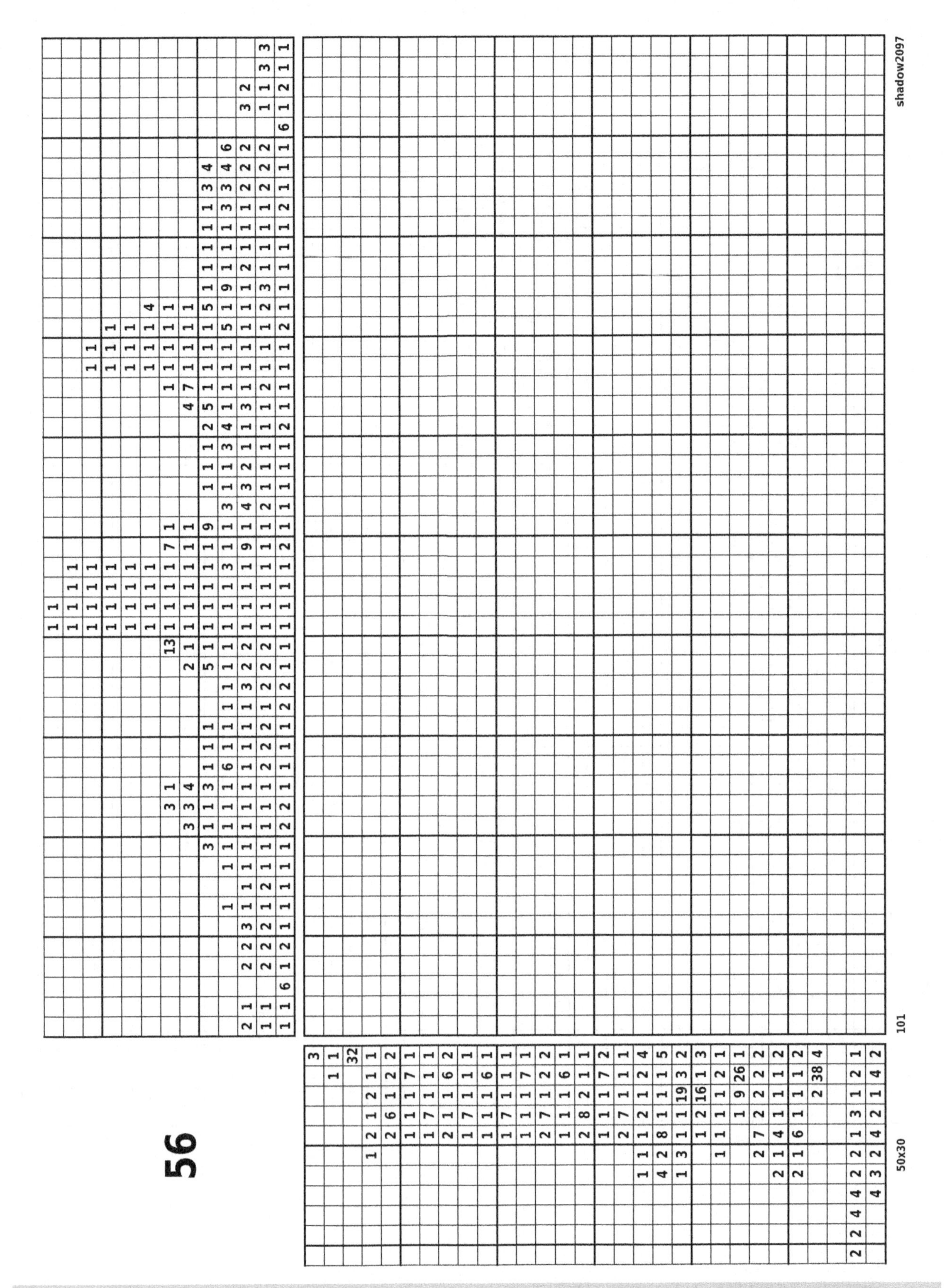
56
50x30
101
shadow2097

57

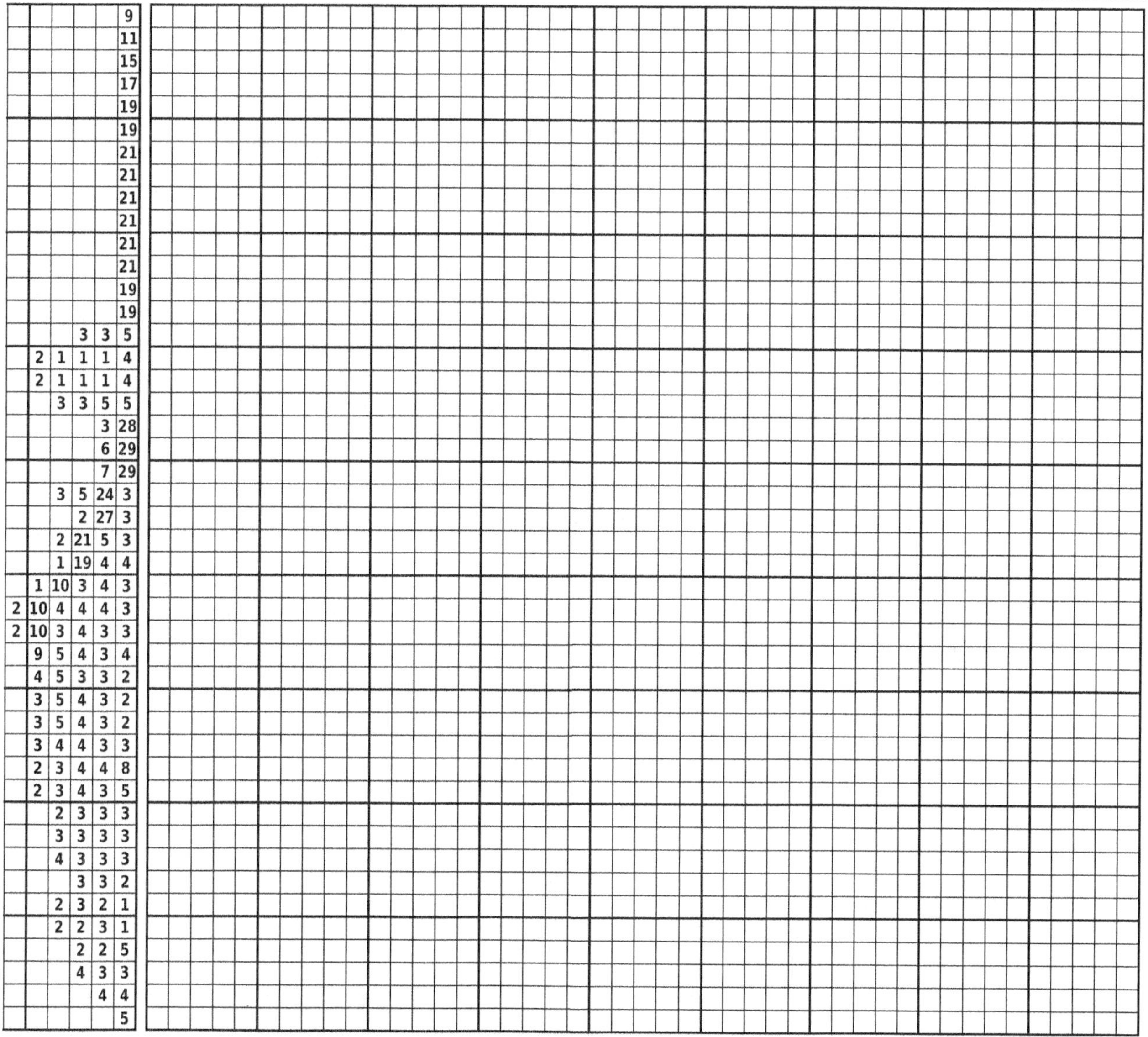

45x45 101 HelleRys

58

Column clues (top to bottom)

Column	Clue
1	1 1 1 1 1 1 1
2	4 1 1 1 1 1 1 2
3	1 1 1 2 1 1 1
4	2 1 1 1 1 1 7 1
5	1 1 1 1 1 1 2 1
6	1 2 1 1 1 2 4
7	2 1 1 2 1 1 4 2
8	4 1 1 1 3 3
9	1 2 2 1
10	1 5 1 3
11	1 3 4 4 2
12	7 1 2 7 3
13	2 11 10 2 1
14	1 2 15 3
15	2 2 1 10 4 2
16	1 2 1 4 3 3 1
17	1 2 4 8 2 4
18	1 2 3 2 2 1 2
19	1 2 3 2 5 1 4 2
20	1 2 4 8 3 3 1
21	1 2 1 2 2 4
22	1 2 1 2 1 2
23	1 2 5 1 4 1 1
24	1 2 5 3 5
25	1 12 2 3
26	1 8 4 2
27	1 10 1 1
28	1 12 5
29	3 3 9 3
30	1 1 4 10 1
31	1 10 10 3 1 1
32	1 12 3 1
33	1 8 2 4
34	1 7 1 2
35	1 7 1 5
36	2 3 3 4 3 2
37	2 2 2 2 8 3 1
38	1 1 2 2 7 1 1
39	2 1 2 8 2 2
40	2 2 2 1 4 5
41	1 1 7 4
42	2 2 3 1 1 2 2 1 1
43	3 3 2 1 1 1 1 1 1 1
44	2 1 1 2 1 1 1 2 2 1 1
45	1 2 1 2 1 1 1 4 1 1
46	1 2 2 2 1 1 1 1 1
47	1 1 2 1 1 1 1 1 1 1 1
48	2 1 2 1 1 1 1 1 1 1 1
49	4 4 1 1 1 1 1 1 1
50	1 1 1 1 1 1 1

Row clues (left to right)

Row	Clue
1	4 5 4
2	1 2 15 3 2 1
3	2 1 3 3 2 1 2
4	1 1 2 1 3 3 1
5	1 3 1 12 1 2 2 1
6	5 15 1 3 4
7	2 2 1 2
8	2 2 1 1
9	2 2 2 1 5
10	2 4 2 1 3 2
11	1 8 2 1 4 2
12	2 1 1 2 2 2 2
13	2 1 1 2 1 1 4 1
14	2 2 1 1 11 1
15	2 4 2 2 6 4
16	14 1 1 2 1 1
17	2 1 1 2 2 4 10
18	5 5 1 2 2 1 8 1
19	3 2 5 1 2 2 11 1
20	2 3 1 2 1 8 4 8
21	1 11 9 1
22	8 1 11 11 2
23	7 14 9
24	7 8 5
25	4 5 7 4 3
26	27 3 2 2 4
27	3 2 2 2 2 10 4 1 1
28	2 1 2 2 2 2 2 2 7 1 1
29	1 1 1 1 1 1 1 2 4 2 2 4
30	1 1 1 1 1 1 1 1 1 2 3
31	2 1 2 2 2 2 2 2 2 1 1 1
32	3 2 2 2 2 2 2 2 1 2 1 9
33	33 1 1
34	1 2 2 2 3 2 3 2 4 2
35	2 3 2 2 2 2 2 2 3 4 9

50x35

101

xiayu

59

Column clues (top to bottom, columns 1–45):

Column	Clues
1	11
2	12
3	12
4	13
5	1 1 1 1 13
6	2 2 3 3 13
7	1 1 4 12 14
8	1 2 8 5 14
9	1 11 3 3 15
10	1 11 1 3 3 15
11	1 11 1 1 1 4 11
12	1 14 4 4 3 10
13	1 15 2 9 9
14	20 9
15	15 2 9
16	10 2 8
17	14 2 8
18	12 2 7
19	1 7 1 7
20	7 1 7
21	9 2 7
22	1 9 1 2 7
23	1 7 1 2 3 7
24	8 2 3 1 4 8
25	1 6 1 1 1 1 2 4 9
26	1 8 1 1 1 1 1 15
27	8 1 1 2 1 15
28	8 1 1 1 1 2 2 12
29	8 1 1 1 3 1 2 11
30	10 1 1 1 1 10
31	1 8 1 1 1 1 1 10
32	6 1 4 1 2 9
33	8 1 1 1 1 8
34	9 1 1 2 1 8
35	1 10 1 1 1 2 7
36	10 1 2 4 1 1 6
37	9 1 1 2 2 2 4
38	11 1 1 1 1 2 2
39	1 11 1 1 3 2
40	1 6 1 2 1 3
41	1 1 8 1 5 3
42	1 9 1 2
43	1 17
44	1 3 3
45	1

Row clues (left to right, rows 1–50):

Row	Clues
1	2 2 2 1
2	2 1 1 3 1 1
3	2 5 6 3 2
4	3 25
5	2 26
6	29 2
7	2 29
8	2 35
9	37
10	12 2 2 2 12
11	10 3 2 1 9
12	12 2 2 7
13	13 1 2 5
14	13 1 4
15	9 1 4 3
16	8 2 2 1 2 2
17	9 1 4 2
18	10 2
19	4 5 6 2
20	2 2 2 1 1 4 2
21	1 2 1 1 1 1 1 2 1 1 2 1
22	1 1 2 1 1 1 1 1 1 2 1 1
23	2 2 1 2 1 3 1 1 1 1 1
24	3 1 1 1 2 1 3 1 1
25	2 2 4 2 1
26	2 2 2
27	3 1 2 2
28	7 2 4 1
29	3 1 3 2
30	1 1 1 4 3 1
31	2 7 2
32	1 2
33	1 2
34	1 3
35	3 6 4
36	5 12
37	7 7
38	8 1 4
39	9 1 4
40	11 4
41	12 6
42	15 8
43	17 11
44	35
45	36
46	36
47	37
48	37
49	38
50	38

45x50

101

xxLadyJ xx

60

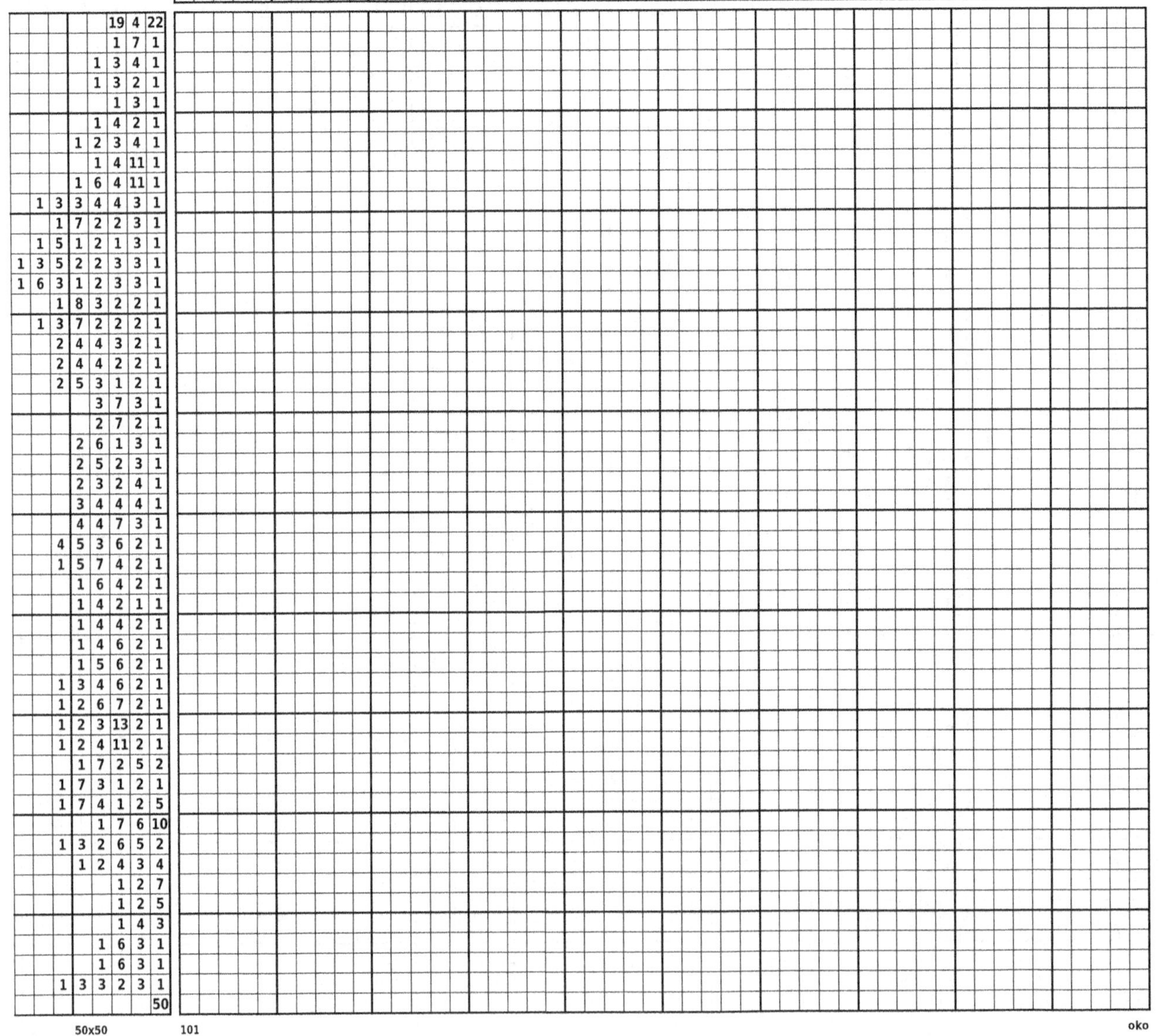

50x50

101

oko

61

Column clues (top to bottom, columns 1–40):

Column	Clues
1	3 36
2	2 1 5 5 15
3	4 1 1 1 11
4	6 4 9
5	4 7 2 2 1 8
6	1 1 5 1 3 3 7
7	2 3 2 4 2 2 2 7
8	4 3 5 3 2 3 3 1 6
9	4 3 12 2 3 2 2 6
10	4 2 7 4 3 2 2 5
11	5 3 1 5 5 4 2 1 5
12	5 2 5 1 4 4 2 1 1
13	6 5 2 3 4 2 1
14	5 1 4 2 3 3 1 2
15	3 3 2 2 1 3 3 2 1
16	2 6 1 2 1 3 1 2 4 2 4
17	1 2 1 2 1 3 1 1 4 2 4
18	2 2 1 2 1 2 1 2 2 2 4
19	2 2 1 2 3 3 3 2 1 2 4
20	2 2 2 2 3 3 3 4 1 2 1
21	1 3 3 3 2 3 3 2
22	1 5 3 3 2 2 3 3 3
23	10 4 3 3 4 3 1
24	10 4 4 1 3 4 2 5
25	11 4 4 2 3 4 2 3
26	11 3 4 2 2 4 2 1
27	11 2 3 3 1 4 2 3 1
28	12 1 3 4 1 3 2 2
29	12 3 3 4 2 6
30	13 3 4 1 3 1 3
31	14 2 3 3 1 4 2
32	15 1 3 1 3 3
33	16 3 1 3 10
34	17 1 1 1 11
35	19 10
36	21 9
37	30 3 9
38	31 2 8
39	31 1 8
40	32 9

Row clues (left to right, rows 1–45):

Row	Clues
1	12 19
2	3 10 2 18
3	4 8 4 18
4	3 7 2 18
5	1 3 4 2 1 18
6	1 2 3 1 1 2 19
7	2 1 5 2 2 19
8	7 2 1 20
9	1 2 2 21
10	1 1 1 1 22
11	2 1 1 1 1 1 1 16
12	3 1 1 2 13
13	2 2 3 2 3 11
14	2 9 8 10
15	2 9 2 9 9
16	1 9 3 8 8
17	1 9 1 1 4 7
18	1 1 3 1 3 3 6
19	1 2 3 18 6
20	2 1 2 2 13 5
21	2 2 6 6 5
22	3 1 7 3 4 4
23	2 3 7 6 4
24	2 9 4 10 4
25	1 10 7 4
26	1 4 3 2 3 4
27	1 2 9 1 4
28	1 12 10 2 4
29	1 12 4 3 4
30	1 8 1 5 4
31	2 2 3 15 3
32	2 3 15 1
33	2 12 10 1
34	2 11 5 1 2
35	3 7 2 1 3
36	3 3 8 3
37	4 20 5 1
38	5 9 8
39	7 1 8
40	9 1 1 1 8
41	11 1 1 1 1 1 8
42	11 1 4 1 1 1 1 1 8
43	11 1 4 1 2 2 9
44	11 4 2 13
45	12 6 18

40x45

101

anawa74

62

50x45

101

hibrahimozer

63

Column clues (top to bottom, columns 1–40):

Column	Clues
1	3
2	6
3	8 3 3
4	2 7 10
5	1 6 4 8
6	2 22
7	2 1 3 5 3 3 1
8	2 3 3 5 2 1
9	2 6 2 2 1
10	11 1 2 7
11	5 3 2 2 8
12	3 3 1 2 3
13	2 3 2 1 2
14	6 3 2 1 2
15	6 4 2 1 1
16	8 1 2 3 1 1
17	10 2 3 1 1
18	10 2 2 1 1 1
19	9 1 2 5
20	14 2 10
21	4 9 1 11
22	2 10 12
23	2 9 13
24	2 15 12 1
25	3 2 8 11 1
26	4 1 1 5 10 1
27	7 1 2 4 10 1
28	10 4 11 1
29	3 4 1 10 1
30	3 6 9 1
31	2 2 3 8 2
32	4 3 7 3
33	3 14 1 2 1
34	3 11 2 2 1
35	3 11 2 2 1
36	13 3 2 1
37	10 3 2 1
38	12 3 1
39	8 3 1
40	6 2 1

Row clues (left to right, rows 1–50):

Row	Clues
1	3
2	4
3	2 2
4	1 1 2
5	2 1 2
6	4 2 2
7	1 2 1 2
8	1 3 1 1 2
9	1 4 3 1 1 2
10	2 11 1 3
11	2 14 1 2
12	1 5 8 1 3
13	2 4 10 2 1 1
14	2 2 17 2
15	2 3 13 1 2
16	2 5 10 2 1
17	4 2 11 2
18	3 2 9 2
19	2 1 2 7 2
20	1 1 2 1 7 1 1
21	3 4 7 2 2
22	5 3 8 1 1
23	4 6 2 2
24	3 3 1 1
25	4 2 4
26	4 2 5
27	5 2 5
28	7 3 6
29	2 1 2 6
30	7 11
31	5 12
32	5 12
33	4 6 14
34	15 14
35	7 2 2 15
36	2 3 9 3
37	2 8 8 3
38	7 8 4
39	1 8 3
40	1 8 2 1
41	1 8 2 2
42	1 8 2 2
43	2 8 2 3
44	2 9 3
45	1 6 2
46	1 7 2
47	1 7 2
48	2 6 2
49	4 5 2
50	30

40x50 101 TNT

64

Column clues (top to bottom):

Column	Clues
1	38 1 1 7
2	1 4 1
3	1 3 1 1
4	1 3 2 1
5	1 7 1 1
6	1 4 1 2 1 1
7	1 6 2 2 2 1
8	1 2 1 2 1 1
9	1 4 2 2 1 1
10	1 3 2 2 2 1 1
11	1 3 2 2 2 2 1
12	1 2 2 2 2 2 1 1
13	1 2 2 2 2 2 1 1
14	1 2 2 2 2 2 1 1
15	1 2 1 1 2 2 2 2 1 1
16	1 2 5 2 2 2 2 1 1
17	1 2 2 2 2 2 2 2 1 1
18	1 2 2 1 2 2 2 4 1
19	1 2 2 1 2 2 2 2 1
20	1 2 2 1 2 2 2 2 1
21	1 2 2 2 2 2 2 2
22	1 2 2 2 2 2 2 1 2
23	1 2 2 1 2 2 2 1 1 1
24	2 2 2 2 2 2 1 1 1
25	2 2 2 2 2 9
26	4 2 2 2 1 1 1
27	1 1 2 2 1 1 1 1
28	1 1 2 2 1 1 1 1
29	1 2 2 2 1 1 1 1
30	1 2 2 2 1 1 1 1
31	1 2 4 2 3 1 1 1
32	1 2 2 1 2 1 1 1 1
33	1 4 2 1 2 1 1 1 1
34	1 2 3 2 1 1 1 1
35	1 2 1 2 2 2 1 1 1
36	1 2 1 2 4 1 1 1
37	1 2 2 2 1 1 1 1 1
38	1 23 1 1 2 1 1 1 1
39	1 1 2 2 2 1 1 1
40	1 23 1 1 1 3 1 1 1
41	1 2 2 1
42	24 8 16

Row clues (left to right):

Row	Clues
1	23 1 16
2	1 3 1
3	1 2 1 1 1 1
4	1 2 1 1 1 1
5	1 2 2 1 1 1
6	1 2 2 1 1 1
7	1 2 2 1 1 1
8	1 2 2 1 1 1
9	1 2 2 1 1 1
10	1 2 2 1 1 1
11	1 2 2 1 1 1
12	1 2 2 1 1 1
13	1 2 2 1 1 1
14	1 2 2 1 1 1
15	1 2 2 1 1 1
16	1 2 1 1 1 1
17	1 3 2 1 1 1
18	1 2 2 2 6 1 1 1
19	1 1 2 6 6 1 1 1
20	1 2 2 2 1 1 1
21	1 4 2 2 1 1 1
22	1 1 2 2 2 1 1 1
23	1 1 2 2 2 1 1 1
24	1 1 2 2 1 1 1 1
25	1 1 2 2 1 5
26	1 4 2 2 2 2 2
27	1 1 2 2 2 2 2 1 1 1
28	1 1 2 2 2 1 2 1
29	1 2 2 2 2 1 2 1
30	1 1 2 2 2 3 1
31	1 1 2 2 2 1
32	1 3 2 2 2 4 1 1
33	1 1 2 2 2 2 2 3 2
34	1 1 2 2 2 2 2 3
35	1 2 2 2 2 2 2 1
36	1 1 2 2 2 2 2 1
37	1 2 2 2 2 2 2 1
38	1 1 2 2 2 2 2 2 1
39	2 2 2 2 2 2 2 1
40	2 2 2 2 2 1 1
41	1 2 2 2 2 2 1
42	4 2 2 2 4 1
43	4 2 1 1 2 1
44	1 5 1 13 1
45	1 8 1 1
46	1 2 19 1
47	1 2 1 1
48	1 2 18 1
49	1 2 1 1
50	20 4 16

42x50

101

Lupos

65

Column clues (left to right, each read top to bottom):

Col	Clue
1	7 1 5
2	1 2 1
3	1 3 1
4	1 2 5 1
5	1 5 6 1
6	1 7 7 1
7	1 17 1
8	1 17 1
9	1 17 1
10	1 16 1
11	16 1
12	17 1
13	18 1
14	20
15	21
16	24
17	26
18	25
19	23
20	21
21	20
22	19
23	8 1
24	4 5 7 6
25	3 6 5 2 3
26	3 6 5 6
27	3 7 4 6
28	3 6 1 2 8
29	3 6 3 1 1 6
30	2 6 4 8
31	3 5 3 8
32	3 6 4 3 3
33	3 5 4 6
34	3 5 3 3
35	2 5 3 8
36	3 5 3 2 5
37	3 4 3 8
38	3 4 3 3
39	3 4 3 10
40	3 5 3 10
41	3 3 3 2 7
42	2 4 3 9
43	2 4 3 3
44	2 3 3 10
45	1 3 3 12
46	3 3 10
47	2 3 1 3
48	3 2 8
49	2 25 9

Row clues (top to bottom, each read left to right):

Row	Clue
1	10 2 1
2	1 2 2
3	1 2 1
4	1 2 2
5	1 2 2
6	1 3 2 1
7	1 3 2 1
8	3 3 1
9	3 3 1
10	4 3 2
11	5 4 3
12	4 4 2 1
13	6 4 3 1
14	2 8 5 3 1
15	3 6 7 3 1
16	2 4 4 7 3 1
17	3 6 1 8 3 1
18	5 7 10 3 1
19	8 9 12 3 1
20	18 10 3 1
21	17 9 3 1
22	17 7 4 1
23	17 6 5 1
24	16 3 5 1
25	17 6 1
26	18 6 1
27	20 4 1
28	22 1 1
29	24 1
30	26 1
31	24 1
32	22 1
33	20 3 3
34	20 4 3 1
35	9 4 3 2 1 3 2
36	7 1 1 3 3 4 6
37	7 10 1 1 4 3 2
38	1 6 1 8 3 4 3 2
39	1 4 1 6 1 3 4 3 2
40	1 3 26
41	1 3 26
42	13 1 27

49x42

101

oko

66

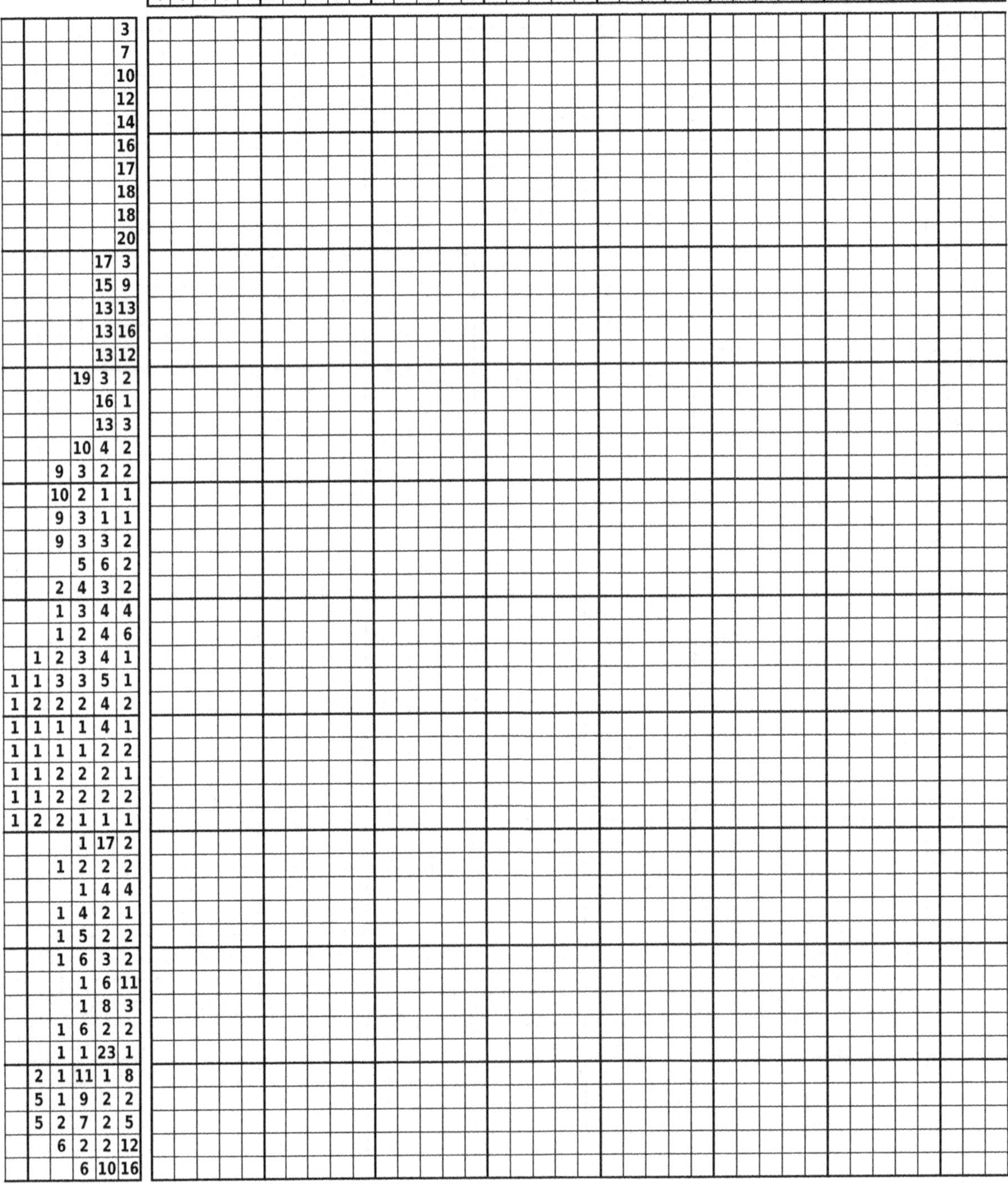

38x50

101

twoznia

67

Column clues (top to bottom):

Column	Clues
1	2 1 1 1
2	2 1 1 2
3	2 1 2 1
4	3 7 1 2
5	6 5 11 1
6	12 2 8 1 6
7	6 7 2 1 4 1 1 2 2
8	3 6 3 4 1 1 1 4
9	3 5 2 1 2 1 1 1 2 3
10	2 4 3 3 1 1 3 2
11	3 4 2 1 2 1 1 2 2
12	3 4 3 6 2 2
13	3 9 7 1 2
14	4 22 2 2
15	12 10 3 2
16	8 19 2 2
17	5 3 11 3 2
18	1 2 2 5 12 2 1
19	2 2 2 8 12 3 2
20	2 2 2 22 2 2
21	4 2 21 3 1
22	3 3 19 3 1
23	2 2 3 3 8 5 1
24	2 3 8 7 7 1
25	5 3 15 1
26	2 3 8 7 7 1
27	3 3 3 3 8 6 1
28	4 3 19 4 1
29	2 2 2 21 3 1
30	2 2 2 7 13 3 1
31	1 2 2 6 12 3 2
32	2 2 5 12 3 2
33	3 4 11 3 2
34	7 18 2 2
35	14 12 3 2
36	31 2
37	5 13 7 1 2 2
38	3 4 2 1 5 1 1 1 2
39	3 3 3 3 2 2 1 2
40	2 3 2 1 2 1 1 1 1 2
41	3 4 3 3 1 1 1 2 2
42	3 5 2 1 3 2 1 1 2 2
43	6 6 3 5 2 1 1 3
44	12 5 5 1 1 4
45	6 5 11 2
46	3 2 7 2
47	2 1 1 2
48	2 2 1 2
49	2 1 1 1
50	

Row clues (left to right):

Row	Clues
1	3 3
2	5 5
3	7 7
4	2 3 3 2
5	2 3 1 3 2
6	3 3 2 1 1 2 3 3
7	2 3 2 2 2 2 3 2
8	2 3 2 3 2 3 2
9	2 3 2 1 1 3 2
10	3 3 13 3 2
11	3 22 3
12	3 4 4 5 2
13	3 3 4 3
14	4 3 4 3
15	3 2 3 3 4 3
16	4 2 5 5 3 3
17	4 2 2 1 1 2 3 3
18	6 1 2 2 2 1 7
19	5 2 3 1 1 2 1 6
20	4 1 3 2 2 3 2 4
21	3 1 5 1 1 5 1 3
22	2 1 5 1 1 5 1 2
23	2 1 5 1 1 5 1 2
24	8 1 5 1 1 5 1 8
25	10 1 5 3 5 12
26	3 1 1 3 1 10 4 1 1 4
27	2 1 1 1 23 1 1 1 3
28	3 1 1 12 11 1 1 4
29	18 18 1
30	1 17 16 2
31	2 3 11 11 3 2
32	2 3 12 12 5
33	4 27 3
34	3 25 4
35	8 23 8
36	5 21 2 2 2
37	8 19 9
38	2 15 3 2
39	2 6 11 8
40	3 6 1 3 5
41	4 2 4 3 9 2
42	3 2 2 5 5 5 2 2
43	3 17 3
44	4 14 2
45	3 3 10 5
46	3 4 6 4 2
47	2 5 4 5 2
48	5 5 2
49	5 5 2
50	15

50x50

101

anawa74

68

								3	3	3		3	2	2	8						5		5						9	2	2			2	3	3								
						3	3	1	1	2	2	2	3	4	3					5	6		9	5					3	3	3	3	2	2	1	1	3	3	4					
					4	1	1	2	2	2	2	3	3	3	9		7	6	5	6	1	22	1	7	5	6	7		9	4	3	3	2	2	2	2	1	1	1					
			6	5	3	2	3	5	3	2	4	1	2	3	3	30	18	16	15	8	1	1	1	8	15	16	18	31	5	3	3	3	4	2	3	4	2	3	4	5	6			
9	11	13	6	4	7	10	12	7	5	5	4	4	4	4	4	4	3	3	4	5	10	9	10	4	4	4	4	4	4	4	4	4	4	5	5	6	12	10	7	4	6	13	11	9

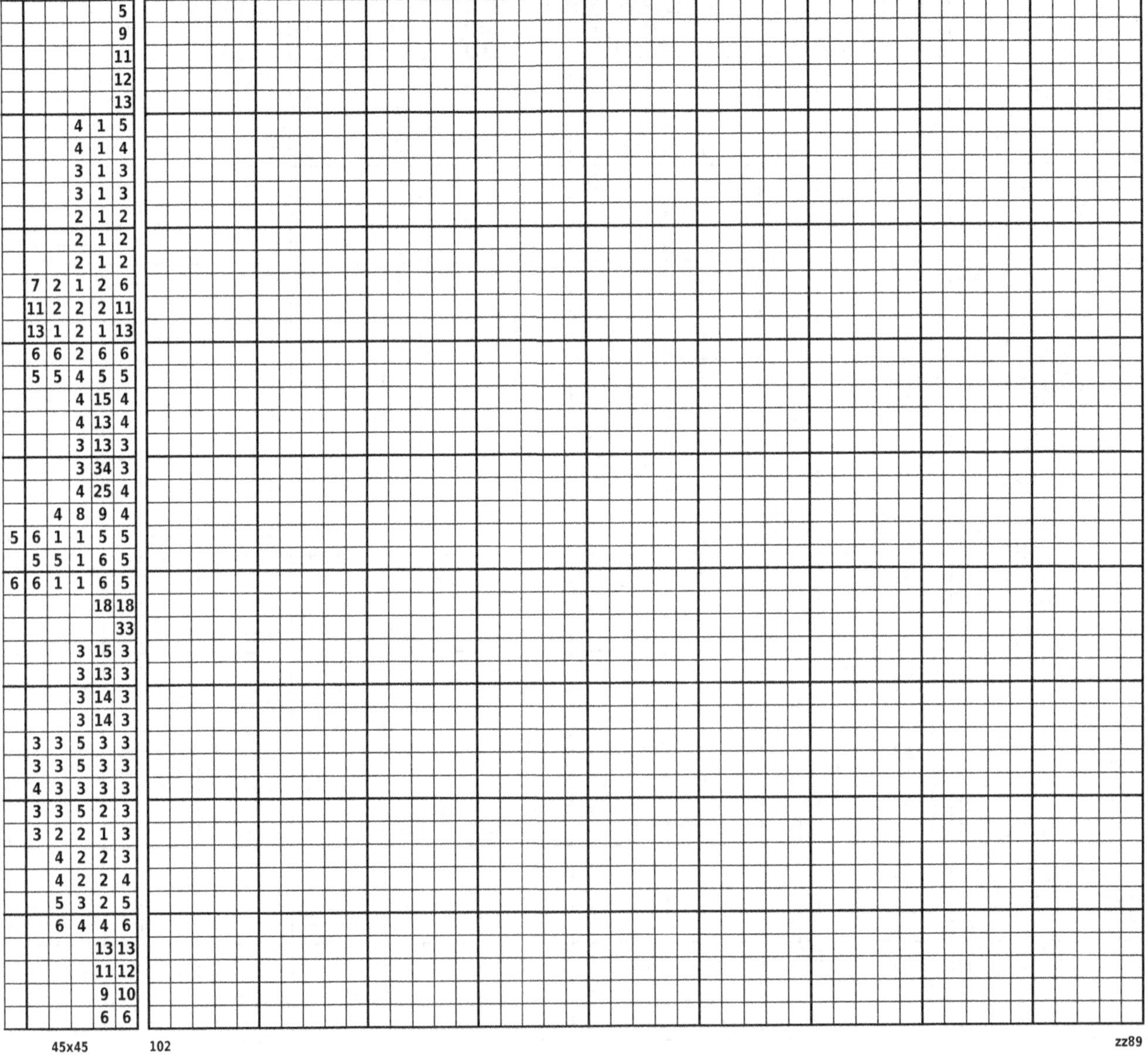

45x45 102 zz89

69

Column	Clues (top to bottom)
1	9 21 1
2	1 2 2 2 1 2 2 5
3	4 2 2 1 2 1 2 2
4	3 1 6 1 2 1 3 1 1
5	1 1 1 5 1 5 4 1 1
6	4 2 3 1 2 4 1
7	3 1 1 2 2 3 3 1 2 2
8	3 3 2 1 3 1 4 2
9	3 5 3 2 1 4 1 1
10	3 2 2 3 3 2 1 2
11	4 3 1 4 2 2 3
12	2 1 3 3 3 2 3 1 2 2
13	2 1 4 2 4 1 2 1 4 2
14	4 3 1 3 2 4 4 1 2
15	2 1 3 1 4 2 4 1 2 1 1 1 1
16	2 1 2 3 4 2 3 5 2
17	2 2 2 2 2 3 2 4 1 2
18	1 1 1 1 3 2 1 3 1 4 5 1 2 1
19	2 2 2 2 2 2 1 3 4 3 1
20	2 2 2 1 2 2 1 1 3 3 4 2
21	2 3 2 5 1 1 3 4 4 1 1
22	2 2 1 3 2 1 2 3 2 3 1 1
23	1 2 2 4 2 1 4 3 1 2
24	2 1 1 5 2 1 3 1 2 1 1 1 2
25	2 2 1 8 1 3 1 1 1 1 1
26	2 1 2 8 1 2 2 2 2
27	2 2 1 6 1 1 2 2 1 4 1
28	1 1 2 3 1 1 2 7 1 3
29	1 1 2 1 1 1 3 17
30	1 1 2 2 1 1 3 1 6 6
31	1 1 2 1 1 1 1 2 1 5 2
32	1 1 2 3 1 2 1 2 7
33	4 2 4 2 3 1 4
34	1 1 3 5 1 2 2 3
35	1 1 1 2 4 2 5
36	2 2 3 3 2 1 4
37	2 1 2 2 4 2 4
38	2 2 2 5 1 2
39	2 3 1 6 1 2
40	1 3 2 2 1 1 1
41	2 3 3 1 1 2
42	1 8 2 1 1
43	3 1 6 1 2
44	3 2 5 2 3
45	8 1 3
46	6 5
47	2 2 3
48	4 7
49	3 8
50	10

Row	Clues (left to right)
1	16
2	2 4 4 4
3	5 5 4
4	4 3 3 1 5
5	4 8 5 3 3
6	2 3 3 4 4 2
7	5 4 2 3 2
8	4 2 3 3 2 3 2
9	3 3 2 3 1 1 4 1
10	3 4 2 4 3 4 2 3
11	3 5 2 6 2 3 1 3
12	1 1 6 2 8 3 2 1 3
13	11 3 9 3 1 7
14	3 2 3 1 5 3 1 7
15	1 1 6 2 2 3 1 1 9
16	3 2 4 2 7 1 2 2 3 2
17	2 2 5 2 4 1 2 3 2 1
18	1 3 2 1 2 4 2 2
19	1 3 3 4 1 7 2
20	2 6 3 5 5 2
21	2 3 4 2 1
22	1 3 3 2 2 3 1
23	1 5 3 5 2 1 2 2 2 1
24	5 10 5 2 1 1
25	4 4 4 5 1 2
26	3 4 2 3 4 1 2
27	2 5 2 5 2 4 2
28	1 5 7 4 3 3 3
29	3 2 2 2 3 5 6
30	1 1 2 2 5 4
31	1 3 4 2 2 3 3
32	3 4 2 5 2
33	2 1 4 3 1 4 1
34	1 3 2 3 2 3 1
35	1 3 4 3 2 3
36	1 2 3 2 1 6
37	1 2 4 5 7
38	2 1 4 1 4 1 4
39	3 1 3 5 5
40	1 1 2 5 1 3
41	1 1 3 2 4 2 3
42	1 3 1 2 2 1
43	1 2 3 4 1
44	1 2 2 3 1 2
45	1 2 3 1 2 1 1 2
46	4 2 1 3 2 2
47	3 1 1 2 4 5 2
48	1 1 2 1 1 1 2 2 4
49	2 2 5 2 1 2 1 4
50	5 24

50x50

102

Andreasss

70

Column clues (top to bottom):

Column	Clues
1	25 13
2	4 1 1 1
3	3 1 1 1 1
4	2 1 1 1 1
5	2 1 1 1 1
6	2 15 10 12
7	1 3 1 1 1 1
8	2 2 1 1 1 1
9	1 2 1 1 1 1
10	1 2 1 1 1 1
11	2 1 1 1 1 1
12	1 2 14 10 7 2 1
13	1 1 3 1 1 2 1
14	1 1 2 1 1 1 3 1
15	2 2 1 1 1 3 1
16	1 1 1 1 1 1 3 1
17	1 1 2 1 1 1 4 3 1
18	1 1 1 1 1 10
19	1 1 2 1 1 14 1
20	1 1 1 1 1 15 2
21	2 2 1 22 2
22	2 1 1 22 4 1
23	2 1 1 22 1
24	1 2 1 23 4 1
25	2 8 32 2 1
26	1 2 1 2 18 2
27	2 1 1 20 2
28	2 1 1 6 13 1
29	2 1 1 1 2 1 7 1
30	1 3 1 1 2 1 1 1
31	1 1 1 1 2 1 1 1
32	1 1 2 1 1 1 1
33	1 1 2 1 1 1 1
34	1 1 2 1 1 1 1
35	3 2 1 1 1 1
36	1 1 16 10 12
37	1 2 1 1 1 1
38	1 1 1 1 1 1
39	2 2 1 1 1 1
40	1 1 1 1 1 1
41	1 2 1 1 1 1
42	2 2 1 1 1 1
43	1 4 1 1 1
44	2 14 10 12
45	2 1 1 1 1
46	2 1 1 1 1
47	2 1 1 1 1
48	2 1 1 1 1
49	2 1 1 1 1
50	16 10 13

Row clues (left to right):

Row	Clues
1	20
2	5 3 1 3 5
3	4 6 1 1 6 4
4	3 3 10 2 3
5	2 3 3 1 3 3 2
6	2 2 3 1 2 3 2
7	2 2 3 1 2 2 2
8	1 2 2 1 1 1 2
9	2 1 1 1 1 1 2
10	1 1 1 1 2 1 2
11	1 1 1 1 1 1 1
12	24 25
13	1 1 1 1 1 1 1
14	1 1 1 1 1 1 1
15	1 1 1 1 1 1 1
16	1 1 1 1 1 1 1
17	1 1 1 1 1 1 1
18	1 1 1 1 1 1 1 1
19	1 1 1 1 2 1 1 1
20	1 1 1 5 1 1 1 1
21	1 1 1 6 2 1 1 1
22	1 1 1 4 4 1 1 1
23	1 1 1 8 1 1 1
24	1 1 1 8 1 1 1
25	1 1 1 7 1 1 1
26	5 5 23 7 5
27	1 1 1 8 1 1 1
28	1 1 1 8 1 1 1
29	1 1 1 10 1 1 1
30	1 1 1 10 1 1 1
31	1 1 1 11 1 1 1
32	1 1 1 11 1 1 1
33	1 1 1 11 1 1 1
34	1 1 1 11 1 1 1
35	1 1 1 11 1 1 1
36	1 1 1 11 1 1 1
37	4 5 23 7 5
38	1 1 1 11 1 1 1
39	1 1 1 11 1 1 1
40	1 1 1 10 1 1 1
41	1 1 1 10 1 1 1
42	1 1 1 11 1 1 1
43	1 1 1 3 5 1 1 1 1
44	1 1 1 2 3 1 2 1 1 1
45	1 1 2 2 2 1 1 1 1
46	1 1 2 1 1 1 1 1 1
47	1 1 1 1 2 1 1 1
48	1 1 2 2 1 1 1
49	1 1 1 2 2 1 1 1
50	5 5 5 2 4 2 6 7 6

50x50

102

mustafademirbas

71

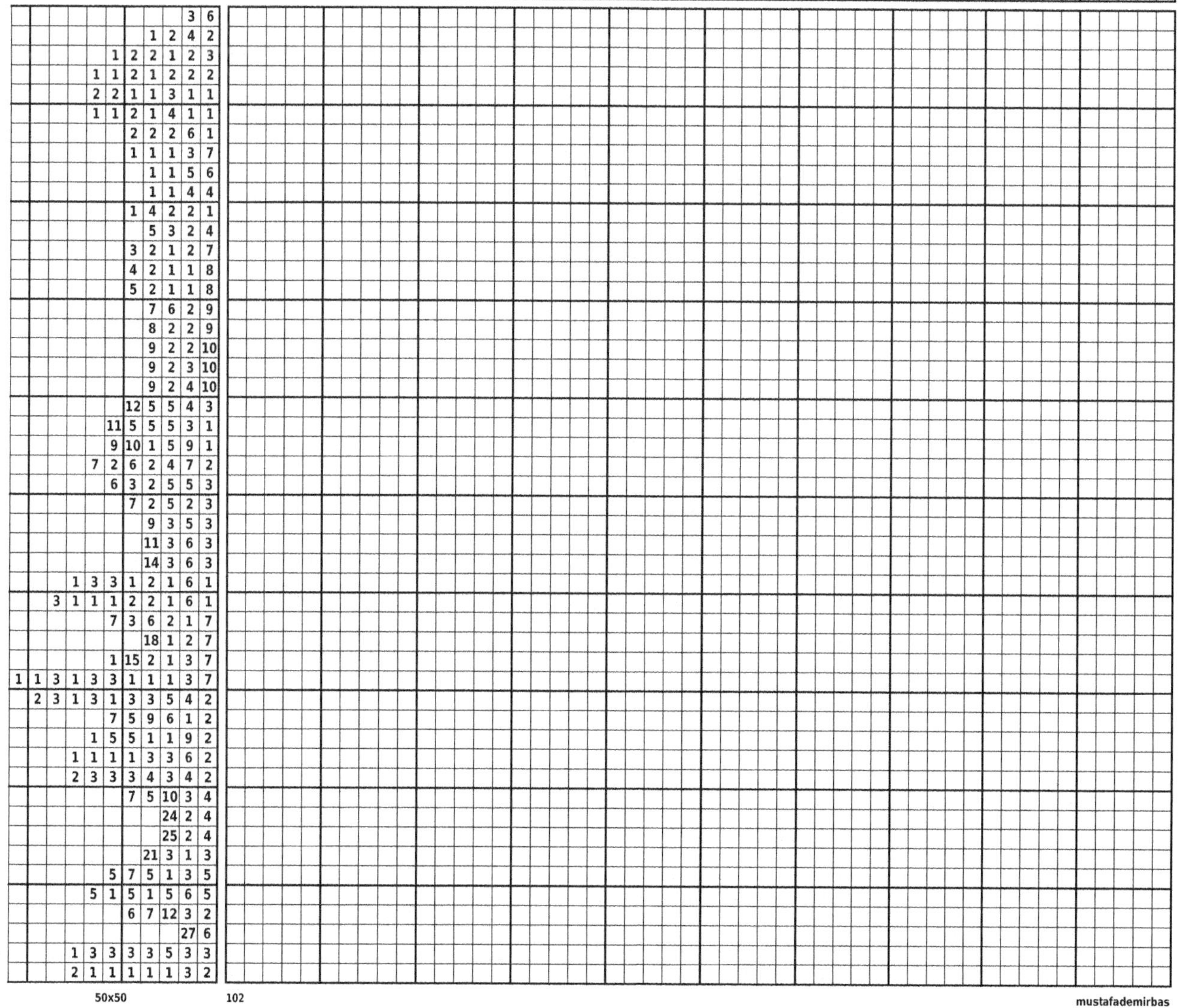

Column clues (top to bottom):

Column	Clues
1	3 6 4
2	1 2 9
3	2 1 2 9
4	6 4 2 11
5	1 1 1 1 9 12
6	1 1 1 1 1 5 2 8 6 2
7	1 1 1 1 1 1 9 3 4 2 4 1 1
8	1 1 1 2 1 18 4 1 1 4 3
9	1 1 1 1 20 4 2 10
10	1 1 1 23 2 3 10
11	1 1 6 16 6 10
12	2 6 14 8 9
13	2 2 14 4 2 9
14	7 5 2 1 1 4 4
15	3 4 4 2 4 1 2
16	2 3 2 8 6 2
17	3 3 1 6 10
18	2 1 3 1 6 2 10
19	2 1 2 2 3 1 2 10
20	2 3 2 3 2 10
21	7 1 4 6 3
22	2 8 1 2
23	2 1 6 4
24	3 10
25	5 7
26	2 2 7 5
27	2 6 1 9 5
28	2 6 11 1
29	3 8 16
30	1 2 5 7 9 1
31	1 2 4 12 3 11
32	3 1 3 15 15
33	3 2 19 8 6
34	4 22 3 9
35	23 14
36	8 22 1
37	9 10 6 5
38	8 9 4 4
39	7 10 2
40	6 4 1
41	5 5
42	4 5
43	3 2
44	3
45	3
46	3
47	2
48	1 7
49	8
50	6

Row clues (left to right):

Row	Clues
1	3 6
2	1 2 4 2
3	1 2 2 1 2 3
4	1 1 2 1 2 2 2
5	2 2 1 1 3 1 1
6	1 1 2 1 4 1 1
7	2 2 2 6 1
8	1 1 1 3 7
9	1 1 5 6
10	1 1 4 4
11	1 4 2 2 1
12	5 3 2 4
13	3 2 1 2 7
14	4 2 1 1 8
15	5 2 1 1 8
16	7 6 2 9
17	8 2 2 9
18	9 2 2 10
19	9 2 3 10
20	9 2 4 10
21	12 5 5 4 3
22	11 5 5 5 3 1
23	9 10 1 5 9 1
24	7 2 6 2 4 7 2
25	6 3 2 5 5 3
26	7 2 5 2 3
27	9 3 5 3
28	11 3 6 3
29	14 3 6 3
30	1 3 3 1 2 1 6 1
31	3 1 1 1 2 2 1 6 1
32	7 3 6 2 1 7
33	18 1 2 7
34	1 15 2 1 3 7
35	1 1 3 1 3 3 1 1 1 3 7
36	2 3 1 3 1 3 3 5 4 2
37	7 5 9 6 1 2
38	1 5 5 1 1 9 2
39	1 1 1 1 3 3 6 2
40	2 3 3 3 4 3 4 2
41	7 5 10 3 4
42	24 2 4
43	25 2 4
44	21 3 1 3
45	5 7 5 1 3 5
46	5 1 5 1 5 6 5
47	6 7 12 3 2
48	27 6
49	1 3 3 3 3 5 3 3
50	2 1 1 1 1 1 3 2

50x50

102

mustafademirbas

72

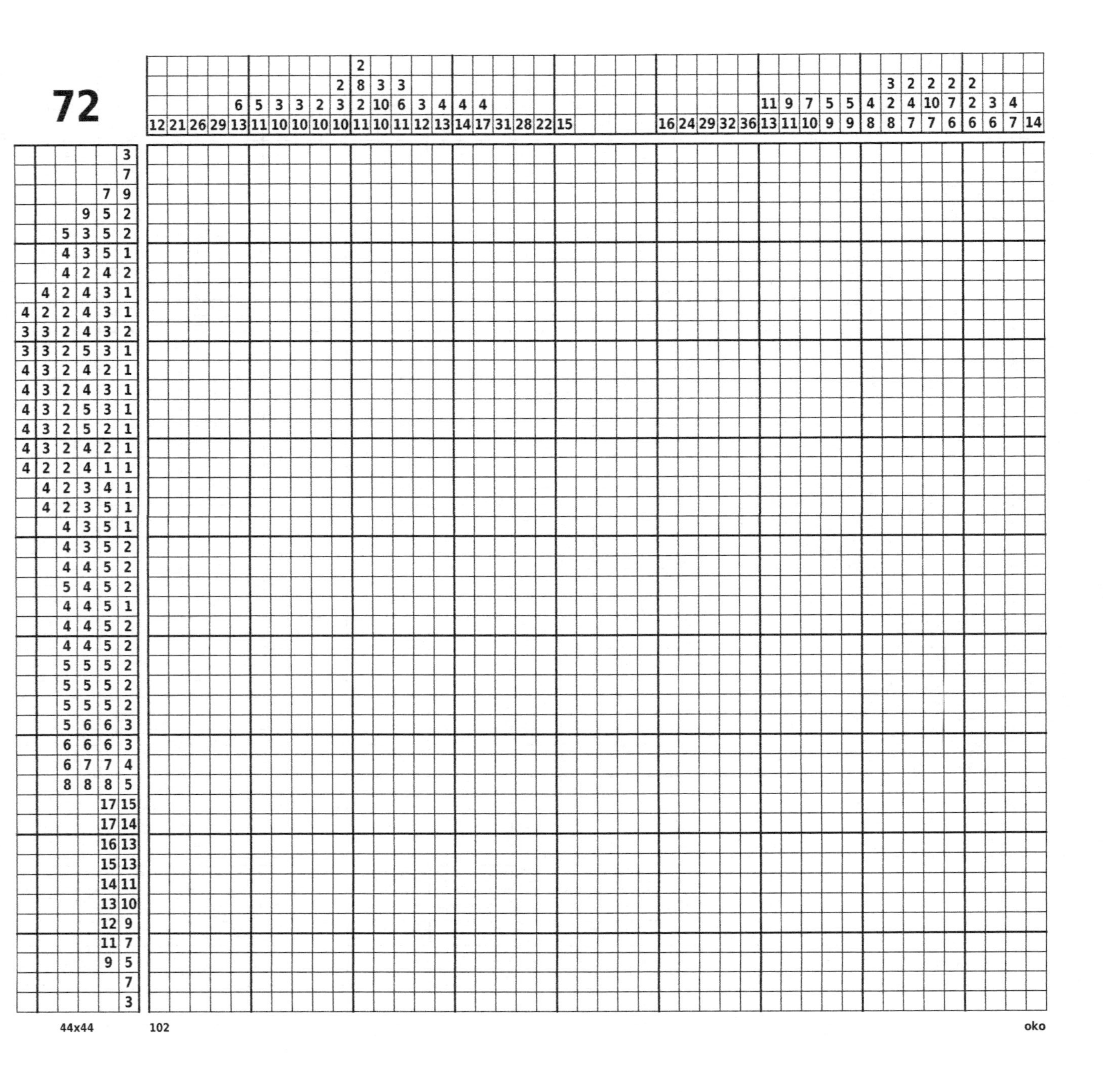

73

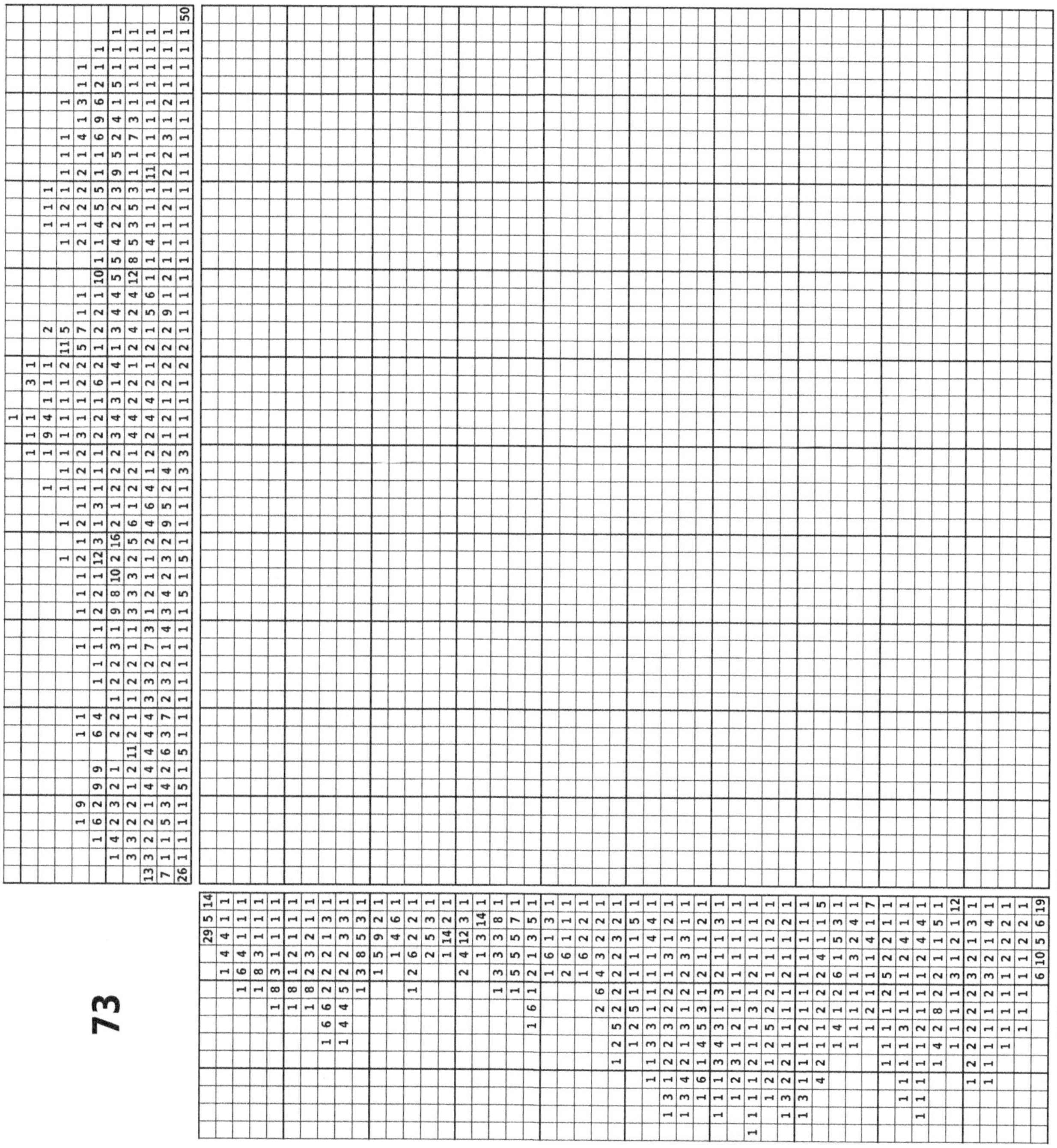

50x50

102

julesc

Column clues (left to right, top to bottom)

1. 13 7 26
2. 1 3 3 1 1
3. 1 4 3 2 1 1
4. 1 6 2 2 2 5 1
5. 9 2 3 2 1 3 1
6. 9 2 1 4 4 5
7. 9 1 2 4 2 1
8. 11 4 6 5
9. 1 6 2 2 4 3 1
10. 1 4 2 1 4 7 1
11. 1 1 3 2 1
12. 1 2 2 3 3 1
13. 1 2 2 2 2 1
14. 1 1 3 1 7 1 1
15. 1 1 1 3 4 1
16. 1 2 9 3 1 3 1
17. 1 2 8 3 2 4 5
18. 1 1 10 3 1 2 1
19. 1 2 12 2 2 1 3 5
20. 1 3 16 5 2 2 1
21. 1 2 1 2 6 4 9 1
22. 1 3 1 1 6 5 1
23. 1 1 1 1 2 2 4 2 1
24. 1 2 1 2 2 1 4 3
25. 1 1 1 2 1 2 1 2 2 3
26. 1 9 1 3 2 3 4 2 1 1
27. 1 1 4 1 1 2 4 4 4 2 1
28. 1 1 1 1 3 2 4 1 1
29. 3 1 1 2 6 1 2 2 2 1
30. 1 1 2 2 2 4 1 1 2 2
31. 11 5 1 1 2 2 2 2
32. 2 5 7 2 3 4 1 2 1
33. 1 2 4 2 5 9 1
34. 1 1 4 4 6 1 1
35. 10 5 12 1 2 1
36. 1 5 8 1 1 1
37. 1 2 1 4 5 4 1 1
38. 1 1 1 4 2 3 1 1 1
39. 1 2 2 5 2 5 1 2 1
40. 1 1 2 5 3 3 1 1 1
41. 1 2 1 9 1 11 2 1
42. 1 1 1 5 1 1 2 1
43. 1 4 6 2 7 1 3 1
44. 1 9 4 3 1 1 1
45. 1 3 6 1 1 1 2 1
46. 1 2 5 1 1 1 1
47. 1 1 1 1 1 1 1
48. 1 1 1 1 1 1
49. 1 1 1 1 1
50. 50

Row clues (top to bottom, left to right)

1. 29 5 14
2. 1 4 4 1 1
3. 1 6 4 1 1 1
4. 1 8 3 1 1 1
5. 1 8 3 1 1 1 1
6. 1 8 1 2 1 1 1
7. 1 8 2 3 2 1 1
8. 1 6 6 2 2 2 1 3 1
9. 1 4 4 5 2 2 3 3 1
10. 1 3 8 5 3 1
11. 1 5 9 2 1
12. 1 4 6 1
13. 1 2 6 2 2 1
14. 2 5 3 1
15. 1 14 2 1
16. 2 4 12 3 1
17. 1 3 14 1
18. 1 3 3 3 8 1
19. 1 5 5 5 7 1
20. 1 6 1 2 1 3 5 1
21. 1 6 1 3 1
22. 2 6 1 1 1
23. 1 6 2 2 1
24. 2 6 4 3 2 2 1
25. 1 2 5 2 2 2 2 3 2 1
26. 1 2 5 1 1 1 1 5 1
27. 1 1 3 3 1 1 1 1 4 4 1
28. 1 3 1 2 2 3 2 2 1 3 1 2 1
29. 1 3 4 2 1 3 1 2 2 3 3 1 1
30. 1 6 1 4 5 3 1 2 1 1 2 1
31. 1 1 1 3 4 3 1 3 2 1 1 3 1
32. 1 2 3 1 2 1 1 1 1 1 1 1
33. 1 1 1 1 2 1 1 3 1 2 1 1 1 1
34. 1 2 1 2 5 2 2 1 1 1 2 1
35. 1 3 2 2 1 1 1 1 2 1 1 2 1
36. 1 3 1 1 1 2 1 1 1 1 1 1 1
37. 4 2 1 1 2 2 2 4 1 1 5
38. 1 4 1 2 6 1 5 3 1
39. 1 1 1 1 1 3 2 4 1
40. 1 2 1 1 1 4 1 7
41. 1 1 1 1 2 5 2 2 1 1
42. 1 1 1 1 3 1 1 1 2 4 1 1
43. 1 1 1 1 1 2 1 1 1 2 4 4 1
44. 1 4 2 8 2 2 1 1 5 1
45. 1 1 1 1 3 1 2 1 12
46. 1 2 2 2 2 2 3 2 1 3 1
47. 1 1 1 1 1 2 3 2 1 4 1
48. 1 1 1 1 1 2 2 2 1
49. 1 1 1 1 1 2 2 1
50. 6 10 5 6 19

74

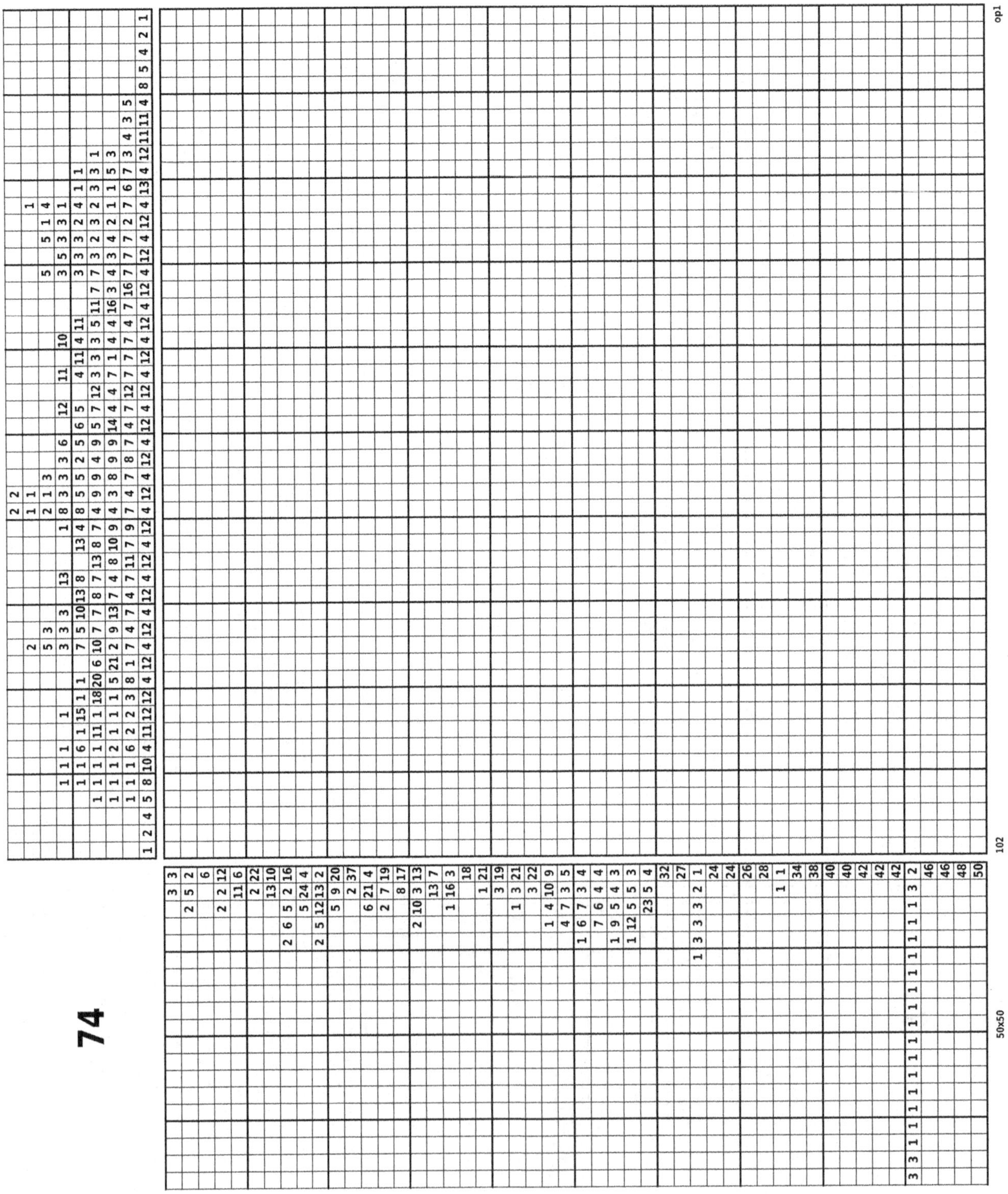

50x50

Row clues:

Row	Clue
1	3 3
2	2 5 2
3	6
4	2 2 12
5	11 6
6	2 22
7	13 10
8	2 6 5 2 16
9	5 24 4
10	2 5 12 13 2
11	5 9 20
12	2 37
13	6 21 4
14	2 7 19
15	8 17
16	2 10 3 13
17	13 7
18	1 16 3
19	18
20	1 21
21	3 19
22	1 3 21
23	3 22
24	1 4 10 9
25	4 7 3 5
26	1 6 7 3 4
27	7 6 4 4
28	1 9 5 4 3
29	1 12 5 5 3
30	23 5 4
31	32
32	27
33	1 3 3 3 2 1
34	24
35	24
36	26
37	28
38	1 1
39	34
40	38
41	40
42	40
43	42
44	42
45	42
46	3 3 1 1 1 1 1 1 1 1 1 1 1 1 1 1 1 3 2
47	46
48	46
49	48
50	50

Column clues:

Column	Clue
1	1
2	2
3	4
4	1 1 1 5
5	1 1 1 1 1 8
6	1 1 1 1 1 10
7	1 6 1 2 6 4
8	1 11 1 2 11
9	1 15 1 1 2 12
10	1 18 1 3 12
11	1 20 5 8 4
12	6 21 1 12
13	2 5 3 7 10 2 7 4
14	3 3 5 7 9 4 12
15	3 10 7 13 7 4
16	13 8 7 4 12
17	13 8 7 4 7 4
18	13 8 11 12
19	13 8 10 7 4
20	1 4 7 9 9 12
21	2 1 2 8 8 4 4 7 4
22	2 1 1 3 5 9 3 4 12
23	3 3 5 9 8 7 4
24	3 2 4 9 8 12
25	6 5 9 9 7 4
26	6 5 14 4 12
27	12 5 7 4 7 4
28	12 4 12 12
29	11 4 3 7 7 4
30	11 3 1 7 12
31	10 4 3 4 7 4
32	11 5 4 4 12
33	11 16 7 4
34	7 3 16 12
35	5 3 3 7 4 7 4
36	5 3 3 3 7 12
37	5 3 3 2 4 7 4
38	1 3 2 3 2 2 12
39	1 4 1 4 2 1 7 4
40	1 3 1 6 13
41	1 3 5 7 4
42	1 3 3 12
43	4 11
44	3 11
45	5 4
46	8
47	5
48	4
49	2
50	1

102

op1

75

50x50

102

Agrippina

76

																		1		1																													
																		2	1	3										1			1																
															1			2	3	3	1							1	1	2		1	6	1		1	1												
															6	1	1	8	8	3	3							3	9	9	1	8	3	4	1	4	3	1											
												1	1	1	3	8	1	2	1	2	2							4	2	3	15	3	4	5	4	6	6	3											
							1	1			1	5	4	4	4	5	14	3	3	2	1	1						11	2	1	2	3	3	5	6	4	4	6	1										
						1	2	1	1	1	4	5	6	5	1	2	3	1	1	1	2	3	1	1	1		4	2	1	1	1	1	1	5	5	2	3	5	2	1	1	1							
				1	1	3	1	2	2	4	5	4	5	5	2	1	1	6	6	4	2	4	11	2	4		2	2	4	6	6	1	1	2	4	3	2	2	5	1	2	1	3	1	1				
	1	1	1	7	1	3	2	4	5	5	4	2	2	1	4	4	6	1	1	1	2	6	4	2	3	5	11	2	1	1	1	6	5	4	1	2	2	1	6	4	3	3	1	7	2	1	1	1	
	2	1	2	3	10	9	6	7	6	5	4	2	2	2	2	1	1	3	2	2	1	2	7	12	7	15	2	1	2	2	3	1	1	2	2	2	2	4	1	7	6	9	10	3	3	1	2	1	
50	1	1	1	1	1	1	1	1	1	1	1	1	1	1	2	3	5	2	2	2	2	8	11	8	8	10	8	2	2	2	2	5	3	2	1	1	1	1	1	1	1	1	1	1	1	1	1	1	50

								50
					1	2	1	1
				1	4	3	3	1
			1	4	3	3	7	1
			1	4	2	10	2	1
		1	2	2	9	2	3	1
		1	2	3	6	3	2	3
		1	2	1	7	4	1	1
		1	1	1	6	5	2	1
	1	3	3	1	5	6	4	1
	1	6	3	1	4	6	4	1
	1	2	10	1	4	6	5	1
	4	1	9	1	5	6	4	1
2	2	3	6	1	5	5	4	1
1	1	3	7	1	4	3	6	1
	1	2	4	6	1	4	8	1
		1	3	6	5	2	14	1
	1	3	6	4	2	11	2	1
	1	4	6	4	2	1	8	1
1	4	6	4	2	2	4	3	1
1	4	4	4	1	1	2	2	1
		1	6	3	3	2	3	1
			1	8	2	3	3	1
			1	15	3	3	2	1
1	2	9	3	2	3	3	3	1
	1	9	2	1	3	3	3	1
		1	3	5	4	3	4	1
				1	2	7	5	1
					1	15	3	1
				1	5	8	3	1
				1	5	7	3	1
						1	6	1
				1	7	4	7	1
		1	3	2	2	2	3	1
	1	2	3	2	2	3	2	1
		1	2	6	4	6	2	1
		1	1	8	4	8	1	1
		1	1	8	4	8	1	1
		1	2	6	4	6	2	1
		1	2	3	6	4	2	1
				1	3	8	3	1
						1	20	1
						1	6	1
					1	4	4	1
			1	3	2	1	3	1
			1	3	1	1	3	1
			1	3	2	2	3	1
			1	3	2	2	3	1
					1	6	6	1
					14	3	3	14

50x50

102

maycanatan

77

Column clues (top to bottom)

Column	Clues
1	37
2	8 1 32
3	3 1 1 5 1 1
4	2 8 1 1
5	2 2 1 1
6	2 2 1 1
7	1 2 1 1 1
8	4 1 1 1 1
9	1 1 2 1 2 1
10	1 1 2 3 1
11	1 1 3 10 1
12	4 3 9 5 1
13	3 1 8 1 11 1 1
14	1 1 2 11 10 1 1
15	1 1 1 14 11 1 1
16	1 1 20 2 4 2
17	1 1 4 18 1 1 4 4
18	3 3 6 3 8 2 8
19	1 1 4 5 3 6 3 4 1
20	1 1 2 4 4 4 1 1
21	3 2 4 6 1 1
22	1 1 2 5 6 1 1
23	1 1 2 3 3 1 3 1 1
24	3 2 5 4 3 1 1
25	1 1 2 7 6 3 1 1
26	1 1 2 4 7 3 1 1
27	3 4 4 3 11 3 1 1
28	1 1 1 4 1 3 1 11 5 1
29	1 1 2 3 4 1 5 1
30	3 6 1 4 12 5 1
31	1 1 4 1 3 1 12 3 1 1
32	1 1 4 3 11 3 1 1
33	3 2 1 2 7 3 1 1
34	1 1 2 3 5 6 3 1 1
35	1 1 6 5 5 3 1 1
36	1 1 4 5 3 1 1
37	1 1 1 2 7 3 2 1
38	3 2 3 5 3 1 1
39	4 2 3 3 3 2 1
40	1 1 2 3 4 1 1
41	1 1 2 5 3 1 1
42	1 1 2 3 4 2 1
43	4 2 3 1 1
44	1 2 6 2 1
45	2 2 2 1
46	2 2 2 1
47	2 8 1 1
48	3 1 1 5 1 1
49	8 1 32
50	37

Row clues (left to right)

Row	Clues
1	28
2	6 1 1 1 1 1 1 6
3	3 28 3
4	2 5 5 2
5	2 2 2 2
6	2 2 2 2
7	1 2 2 1
8	4 4
9	1 1 1 1
10	3 3
11	1 1 2 1 1
12	1 1 4 3 1 1
13	3 4 1 3 3
14	2 1 2 1 2 2 1 2
15	4 3 2 2 2 4
16	1 1 2 3 2 2 1 1
17	3 5 3 3
18	1 1 7 1 1 1
19	3 8 1 8 3
20	2 9 3 3 3 2
21	2 7 3 3 6 2 2
22	2 2 5 1 1 6 1 2
23	2 2 5 4 2 2
24	2 3 5 2 2 5 2
25	2 6 4 2 5 2
26	2 6 4 2 2 2 2 2
27	2 8 4 1 2 2 2 2 2
28	2 1 4 3 3 2 3 2 2 2 2
29	2 2 4 2 3 2 3 3 2 2 2
30	2 2 4 3 2 2 3 3 3 2 2
31	2 3 4 3 1 2 3 4 3 2 2
32	2 4 4 2 3 4 4 3 1 2
33	2 4 4 2 4 5 4 2 1 2
34	2 5 3 1 4 6 3 2 1 2
35	2 5 3 5 6 3 1 2 2
36	2 5 3 8 6 2 1 3 2
37	2 1 3 3 5 6 1 4 2
38	2 1 3 6 4 5 5 5
39	2 1 3 6 3 7 2 2
40	2 2 3 21 2 2
41	2 2 2 17 3 2
42	2 2 3 12 4 2
43	9 2 3 3 3 2
44	2 6 2 8 7 2
45	2 3 3 2
46	2 3 2 2
47	2 3 2
48	2 3 2
49	2 3 2
50	50

50x50

103

maycanatan

78

					2	3	3	3												2	3	2	2											
			1	1	1	1	1	1			3					4	4			1	2	2	2	3	2									
		1	4	3	1	2	1	1	2	2	4	2	2			1	4	1	2	2	2	2	3	4	1	2	3							
2	5	2	2	4	6	3	2	3	1	3	1	4	6	2		9	1	6	2	8	2	2	1	2	4	4	15		5	3	1			
8	20	4	3	1	2	2	5	4	1	11	14	21	20	28	31	5	3	5	4	4	3	3	1	1	2	21	4	31	24	22	6	5	4	3

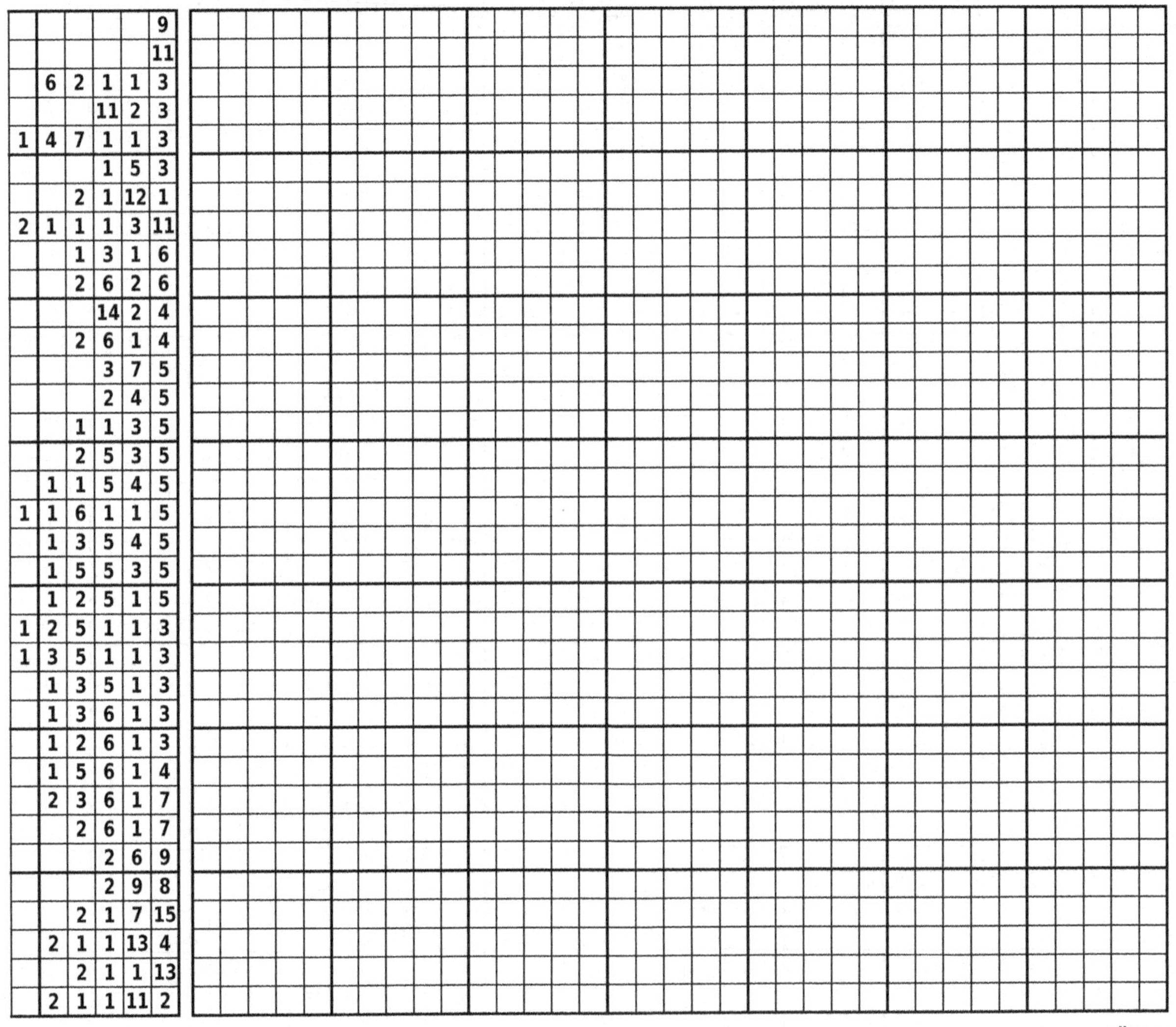

35x35 102 elimaor

79

Column clues (top to bottom, columns 1–50, left to right):

Column	Clues
1	11 24 2 2 1
2	1 5 2 2 1
3	1 2 2 4 7 2 3 1
4	1 1 1 11 3 4 2 3 1
5	1 1 4 4 3 4 1
6	1 2 6 2 6 1
7	1 3 4 3 5 1
8	1 6 4 4 6 1
9	1 10 3 5 1
10	12 4 6
11	5 6 5 5
12	5 6 2 2 5
13	4 6 3 2 4
14	4 5 2 2 1 2 4
15	3 1 3 2 4 2 3
16	3 2 4 4 2 3 3
17	4 3 1 3 1 5 2 1
18	4 2 2 1 3 1 1
19	5 2 3 2 2 1 2 1
20	4 3 2 3 1 2 2 2
21	5 5 2 2 2 1 2 3
22	1 5 6 2 3 1 1 2 1 1
23	1 5 11 2 2 3 1 1 2 1
24	1 3 9 2 4 2 1 1 2 1
25	1 2 10 2 2 1 1 2 1
26	1 2 11 2 2 1 10
27	1 1 2 9 2 4 1 2 1
28	1 2 2 9 3 3 3 1
29	1 2 2 7 2 1 1 1 5 1
30	1 2 2 8 1 1 2 1 1 1
31	1 2 2 8 2 2 2 1 5 1
32	1 1 2 2 8 3 6 1 1
33	1 2 3 13 6 1 5 1
34	1 3 3 10 7 1 1 1 1 1
35	1 3 3 11 7 2 1 1 2 1
36	1 3 3 19 3 1 1
37	4 3 3 17 1 1 1
38	1 2 3 4 16 1 1 5 1
39	1 1 4 4 12 2 1 1 1
40	1 4 17 3 1 1
41	1 4 14 5 1 1 1
42	1 5 11 4 1 5 1
43	1 14 4 1 1 1
44	1 10 4 1 1
45	1 1 1 4 1 5 1
46	1 3 2 5 5 1 1 1 1
47	3 2 2 7 5 1 1 2 1
48	1 2 13 1 1
49	1 1 11 1 1
50	25 8 14

Row clues (left to right, rows 1–50, top to bottom):

Row	Clues
1	9 8 29
2	1 12 1 1 1
3	1 1 15 2 2 1
4	1 6 8 1 3 2 1
5	1 5 8 2 1
6	1 4 7 1 1
7	1 3 1 2 1
8	1 1 3 2 11 1
9	1 4 3 15 1
10	1 5 6 6 1
11	1 4 5 14 5 1
12	3 2 6 17 4 1
13	1 1 1 5 10 5 4 1 1
14	1 1 2 4 3 11 5 3 3 1
15	1 1 1 3 1 12 4 3 2 3
16	3 1 4 13 3 2 1
17	1 2 2 2 14 3 3 1
18	1 1 1 2 2 12 3 2 1
19	1 1 2 2 3 12 2 2 1
20	1 1 1 2 2 11 5 1
21	1 2 5 2 1 9 5 1
22	1 1 2 2 1 2 13 1
23	1 1 1 1 2 2 2 12 1
24	1 1 2 1 3 2 12 2 1
25	1 1 1 2 2 2 12 3 1
26	1 2 1 2 5 2 2 10 4
27	1 1 1 3 2 2 2 8 4
28	1 2 1 1 3 2 8 5
29	1 1 2 2 5 3 8 4
30	1 1 1 4 2 2 9 4
31	1 1 1 1 1 2 9 3
32	1 1 2 3 5 8 3
33	1 1 2 1 3 5 8 4
34	1 2 1 1 3 2 8 1 5
35	1 1 1 1 2 9 11
36	1 1 2 2 5 3 12
37	1 2 1 1 1 1 10 1
38	1 1 1 2 2 2 6 1
39	1 2 2 2 2 3 1
40	1 2 1 1 6 3 1
41	2 2 2 3 1
42	2 2 26
43	1 5 5 1
44	7 2 1 1 1 3 3 3 3 1
45	8 2 1 1 1 1 1 1 1 1
46	2 7 2 1 1 3 1 1 1 1 1
47	2 8 3 2 1 1 1 1 1 1 1 1 1 1
48	1 10 3 1 1 1 3 1 1 3 1
49	7 2 1 1
50	50

50x50

103

fordpre

80

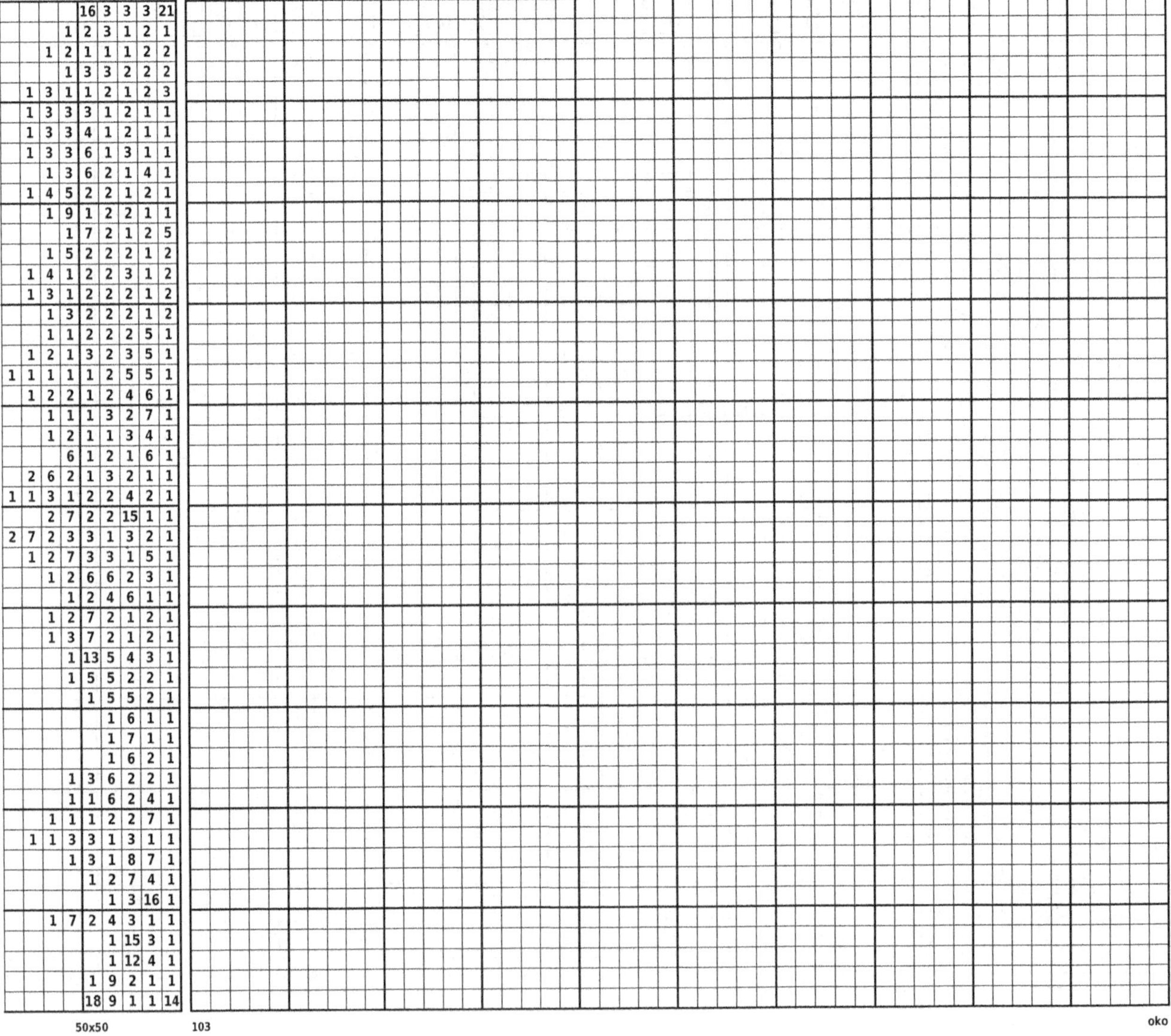

50x50

103

oko

81

Column clues (columns 1–40, top to bottom):

Column	Clues
1	12 8 3 1 1 2 2
2	9 14 3 2 2 5
3	7 23 2 2 1 2
4	5 13 7 2 2 3 1
5	4 3 11 6 1 4 1 1
6	3 3 15 10 1 1
7	2 13 2 2 10 1 1
8	1 3 7 5 13 1 1
9	1 2 5 2 4 12 3 1
10	2 3 4 3 7 2 1
11	1 3 3 3 8 1 1
12	1 4 3 1 2 5 3
13	4 4 1 1 2 2 2
14	5 5 1 1 2 1 4
15	7 4 1 2 1 1 2
16	1 7 4 1 1 1 1 1 1
17	8 2 1 3 1 1 1 1
18	10 2 1 3 1 1 1 1
19	11 3 1 1 1 1 1 1
20	13 3 1 1 1 1 2
21	11 3 1 1 1 2
22	8 2 1 2 1 2 1 4
23	1 12 1 4 1 1 1 1 1 1
24	2 9 1 10 1 1 2 2 1
25	2 8 4 3 7 4 2 2 3
26	12 1 1 1 1 2 2 8
27	8 3 1 1 1 2 6 1
28	1 5 4 1 1 6 3
29	2 1 6 2 3 1 4 12
30	5 3 3 3 6 8 4
31	8 1 4 1 8 2 3 2
32	12 7 7 2 8 1
33	45 2
34	1 26 6 9 1
35	2 24 6 2 3 4 1
36	3 28 2 2 1 2 3
37	5 25 2 2 2 2 1
38	7 17 2 1 1 3 1
39	10 12 12 1
40	14 15 1

Row clues (rows 1–50, left to right):

Row	Clues
1	9 4 2 4 7
2	7 7 2 5 6
3	6 2 7 2 5 5
4	5 3 9 1 7 4
5	4 3 13 6 4
6	3 3 2 15 6 3
7	3 3 3 17 6 3
8	2 26 7 2
9	2 2 4 19 6 2
10	1 2 4 6 4 8 6 2
11	1 10 1 3 5 10 1
12	1 6 2 1 5 1 7 1
13	7 1 1 1 1 7 1
14	9 1 1 1 6 1
15	6 1 1 1 8
16	7 2 2 1 1 1 9
17	7 1 9 1 2 10
18	6 2 10 15
19	6 2 4 6 12
20	6 2 2 3 2 8
21	3 3 1 2 8
22	3 4 4 2 4 7
23	3 1 1 2 3 3 3 7
24	2 2 1 2 2 2 8
25	4 1 3 8
26	4 2 4 9
27	1 3 1 3 8 1
28	1 3 1 1 1 2 1 4 2 1
29	6 2 3 3 3 1
30	3 4 1 1 2 4 4 4 1
31	3 4 1 4 2 5 2
32	1 1 5 2 5 2
33	1 1 5 1 1 3 4 2
34	3 6 11 2 4 1 2
35	1 1 7 1 2 2 6 2
36	1 1 7 4 1 1 2 1 3
37	3 8 1 1 2 1 1 1 2
38	1 1 7 2 1 2 2 3 2
39	2 1 3 2 2 4 3 1 3
40	2 1 1 2 2 5 4 1 2
41	1 2 2 2 13 9
42	4 2 2 9 4 2
43	17 9 2
44	17 2
45	2 1 6
46	1 2 3
47	1 1 1 2 2
48	1 2 3 1 4 6
49	11 1 6 1
50	11 2 1 2 4

40x50 103 Glucklich

82

1	2	3	4	5	6	7	8	9	10	11	12	13	14	15	16	17	18	19	20	21	22	23	24	25	26	27	28	29	30
																			5										
																		5	1										
						1	1		1	1	1	1	1	8	7		6	2	2	8									
					1	6	2	1	2	4	6	5	6	2	2	7	2	2	3	2			1						
		1	1		7	2	3	2	4	2	1	2	4	2	2	2	3	2	1	1	16	14	12	1		1			
	1	5	7	1	2	3	1	4	1	6	5	5	7	5	1	5	1	1	2	3	2	1	3	7	1	1	1	1	
	2	2	4	8	2	1	1	2	2	2	2	2	2	1	4	2	2	4	2	1	3	1	2	1	1	3	1	2	
	2	3	2	3	2	2	3	4	2	1	2	3	1	1	2	3	4	1	2	3	2	2	1	1	1	2	1	1	
	5	2	1	2	1	1	1	1	1	1	1	2	1	1	2	1	2	1	2	7	2	1	12	2	1	14	2	2	
50	1	1	1	7	2	1	1	1	1	1	1	1	1	1	1	1	2	3	1	1	1	1	1	26	25	1	1	1	50

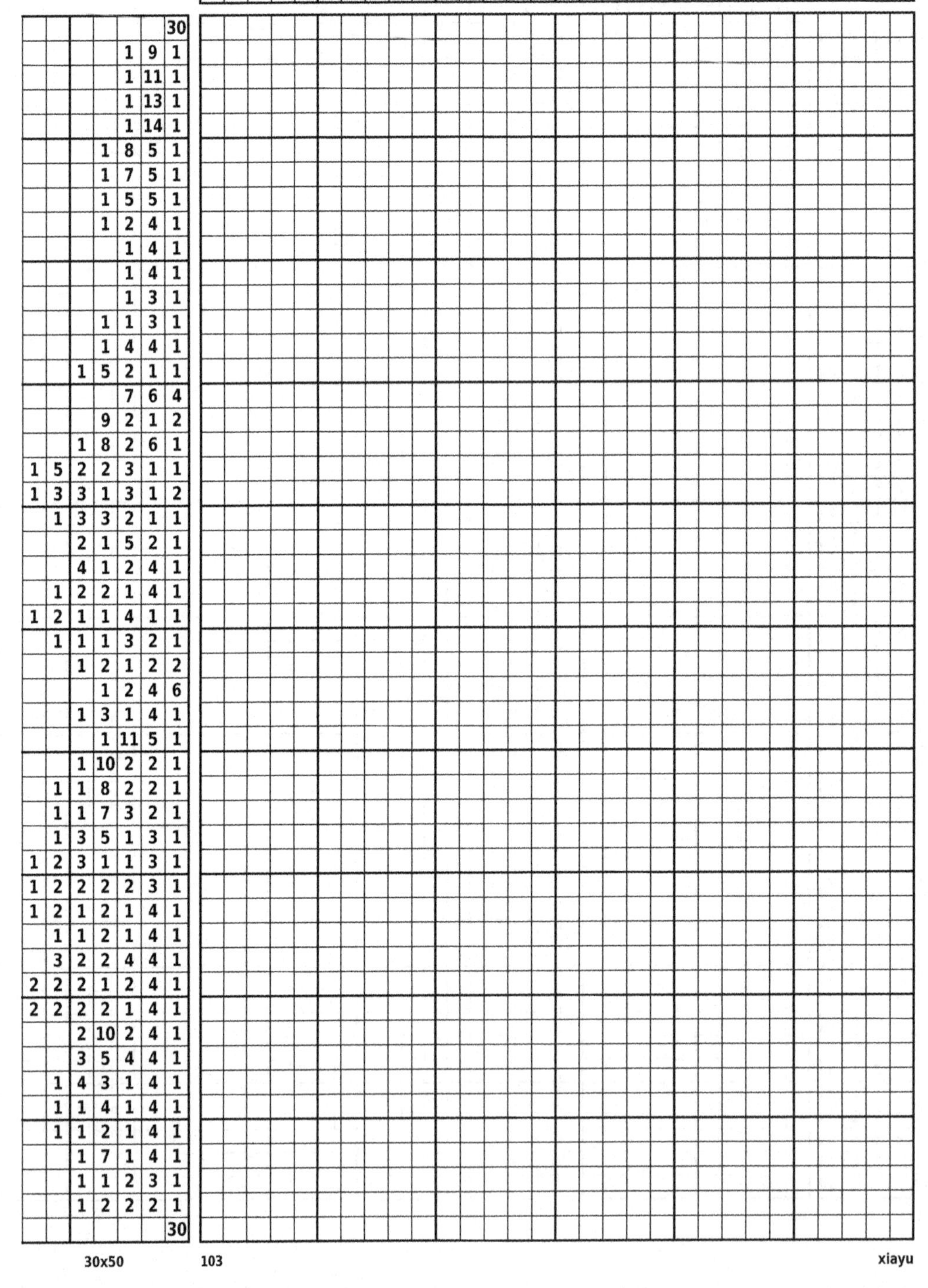

						30
				1	9	1
				1	11	1
				1	13	1
				1	14	1
			1	8	5	1
			1	7	5	1
			1	5	5	1
			1	2	4	1
				1	4	1
				1	4	1
				1	3	1
			1	1	3	1
			1	4	4	1
		1	5	2	1	1
				7	6	4
			9	2	1	2
		1	8	2	6	1
1	5	2	2	3	1	1
1	3	3	1	3	1	2
	1	3	3	2	1	1
		2	1	5	2	1
		4	1	2	4	1
	1	2	2	1	4	1
1	2	1	1	4	1	1
	1	1	1	3	2	1
		1	2	1	2	2
			1	2	4	6
		1	3	1	4	1
			1	11	5	1
		1	10	2	2	1
	1	1	8	2	2	1
	1	1	7	3	2	1
	1	3	5	1	3	1
1	2	3	1	1	3	1
1	2	2	2	2	3	1
1	2	1	2	1	4	1
	1	1	2	1	4	1
	3	2	2	4	4	1
2	2	2	1	2	4	1
2	2	2	2	1	4	1
		2	10	2	4	1
		3	5	4	4	1
	1	4	3	1	4	1
	1	1	4	1	4	1
	1	1	2	1	4	1
		1	7	1	4	1
		1	1	2	3	1
		1	2	2	2	1
						30

30x50

103

xiayu

83

Column	Clues
1	50
2	1 1
3	1 2 37
4	1 1 5 15
5	1 2 3 17 1
6	1 2 2 13 2 1
7	1 2 1 9 11
8	1 2 2 5 12
9	1 1 1 9 1
10	1 1 1 1 6 1 1
11	3 2 1 6 2 1
12	5 1 5 2 1
13	4 1 4 2 1
14	3 1 3 2 2
15	3 2 3 2
16	4 2 3 2
17	1 2 3 2 1 1 1
18	1 2 3 2 1 1 1
19	1 2 3 4 4 2 1 3
20	1 3 2 6 6 3 1 3
21	1 4 3 2 2 2 3 11
22	1 44
23	1 38 2 1 1
24	1 4 3 3 2 2 4 2 1 1
25	1 2 2 2 6 2 2 2
26	1 1 3 3 4 2 2
27	1 1 3 2 1
28	1 2 2 3 1 1
29	1 2 2 3 2 1
30	1 2 1 3 2 1
31	3 1 3 2 1
32	3 1 3 2 2 1
33	3 1 2 3 2 1
34	3 1 3 3 1 1
35	3 1 4 3 2 1
36	3 1 3 4 5 1
37	2 2 3 6 4 1
38	2 1 1 2 6 1
39	3 3 3 2 3 3 1
40	4 4 3 7 3 1
41	4 12 5 4 1
42	2 2 7 11
43	2 3 14
44	2 1 1 8 3 1
45	2 1 2 8 5 1
46	2 1 1 9 5 1
47	1 2 8 6 1
48	1 1 5 7 1
49	1 1
50	50

Row	Clues
1	50
2	1 6 16 1
3	1 14 9 3 1
4	1 4 2 6 7 3 2 1
5	1 4 2 2 3 6 1
6	1 1 2 1 2 1 1
7	1 5 4 5 3 1
8	1 12 2 4 7 2 1
9	1 3 11 3 1
10	1 2 6 2 1
11	1 2 4 3 1
12	1 2 2 2 1
13	1 1 2 3 2 1
14	1 2 2 5 2 1
15	1 2 2 3 4 2 1
16	1 1 2 3 3 2 1
17	1 1 2 3 3 2 1
18	1 1 2 3 2 1
19	1 1 2 3 2 1
20	1 1 6 2 1
21	1 1 7 2 1
22	1 1 6 3 1
23	1 1 2 2 2 1
24	1 1 2 2 3 1
25	1 1 1 5 2 1
26	1 1 1 6 2 1 1
27	1 1 2 5 2 2 1
28	1 1 2 2 2 2 2 1
29	1 1 3 2 2 2 2 1
30	1 1 2 5 2 3 1
31	1 1 3 6 2 2 1
32	1 1 3 1 5 2 3 1
33	1 1 3 2 2 2 2 3 1
34	1 1 3 3 2 2 2 4 1
35	1 1 3 3 5 1 2 3 1
36	1 4 4 4 2 3 3 1
37	1 4 4 3 2 3 4 1
38	1 4 4 3 3 2 4 1 1
39	1 4 4 4 5 3 1 1
40	1 9 5 4 3 2 1
41	1 8 5 4 3 2 1
42	1 3 3 5 5 2 3 1
43	1 7 6 2 5 4 1
44	1 10 4 2 7 4 4 1
45	1 3 3 6 13 3 3 1
46	1 3 2 15 6 1
47	1 3 2 2 4 1
48	1 2 2 8 3 1
49	1 2 2 3 4 2 2 1
50	50

50x50

103

carm53

84

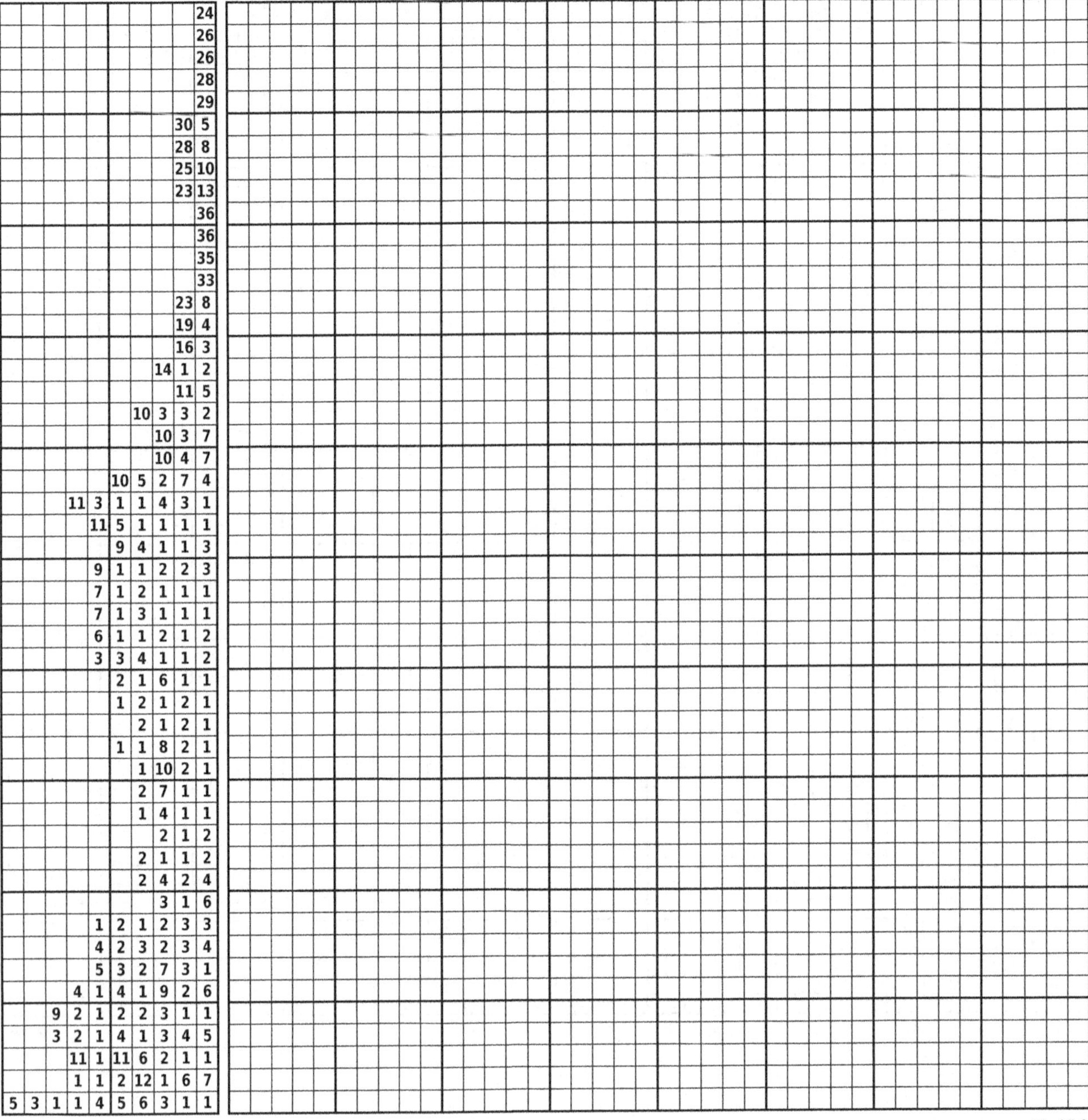

40x50

103

Glucklich

85

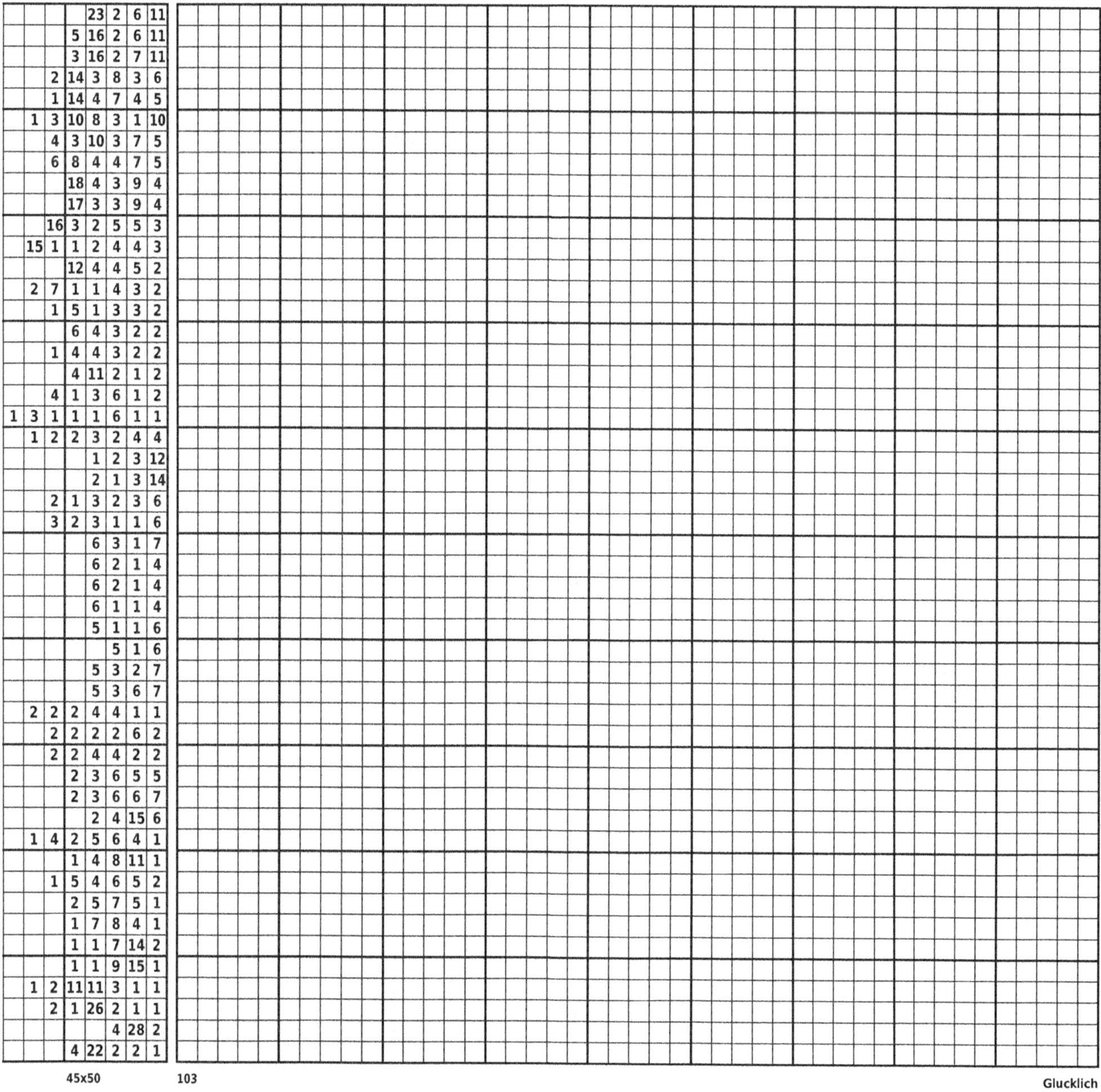

45x50 103 Glucklich

86

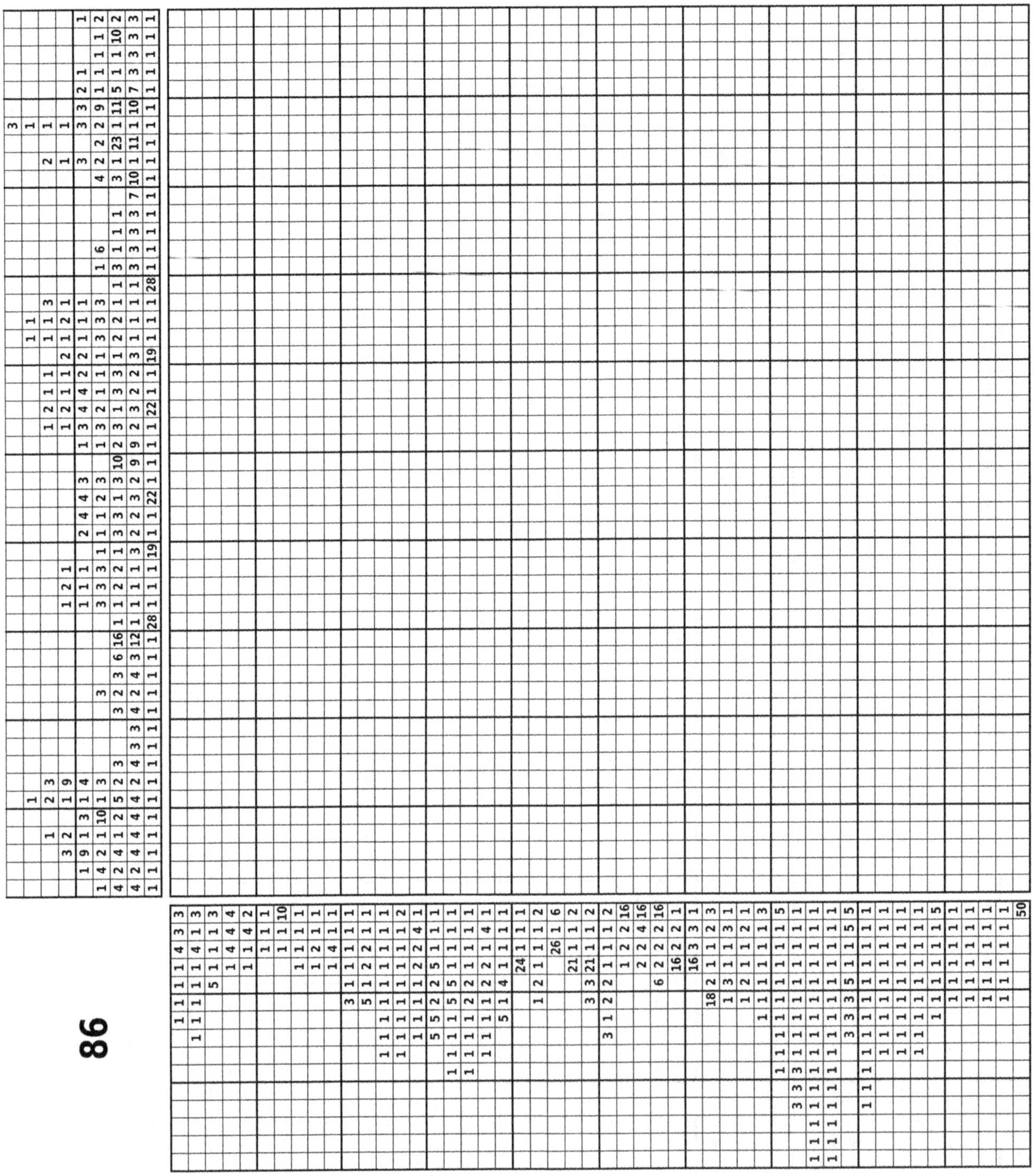

50x50 · 103 · stumpy

Column clues (left to right, each top to bottom)

Col	Clues
1	1 4 4 1
2	1 4 2 2 1
3	3 9 2 4 4 1
4	1 2 1 1 1 4 1
5	3 10 2 4 1
6	1 2 1 1 1 5 4 1
7	3 9 4 3 2 2 1
8	3 4 1
9	3 1
10	3 1
11	3 4 1
12	3 2 2 1
13	3 4 1
14	6 3 1
15	16 12 1
16	1 1 28
17	1 1 3 1 1 1
18	2 1 3 2 1 1
19	1 1 3 2 1 1
20	1 1 3 19
21	2 1 3 2 1
22	4 1 3 2 1
23	4 2 1 3 22
24	3 3 3 2 1
25	10 9 1
26	1 1 2 9 1
27	1 1 3 3 3 2 1
28	2 2 4 2 1 3 22
29	1 1 4 1 3 2 1
30	1 1 2 1 3 2 1
31	2 2 1 1 3 19
32	1 1 1 1 3 2 1 1
33	1 1 2 1 3 2 1 1
34	3 1 1 3 1 1 1
35	1 1 28
36	1 3 3 1
37	6 1 3 1
38	1 3 1
39	1 3 1
40	7 1
41	4 3 10 1
42	2 1 3 2 1 1 1
43	2 23 11 1
44	3 1 1 1 3 2 1 1 1
45	3 9 11 10 1
46	2 1 5 7 1
47	1 1 1 3 1
48	1 1 3 1
49	1 10 3 1
50	1 2 2 3 1

Row clues (top to bottom, each left to right)

Row	Clues
1	1 1 1 1 4 3 3
2	1 1 1 1 1 4 1 3
3	5 1 1 1 3
4	1 4 4 4
5	1 1 4 2
6	1 1 1
7	1 1 10
8	1 1 1 1
9	1 2 1 1
10	1 4 1 1
11	3 1 1 1 1 1
12	5 1 2 2 1 1
13	1 1 1 1 1 1 1 1 1
14	1 1 1 1 1 1 1 1 2
15	1 1 1 1 2 2 4 1
16	5 5 2 2 5 1 1 1
17	1 1 1 1 5 5 1 1 1 1
18	1 1 1 1 2 2 1 1 1 1
19	1 1 1 1 2 2 1 4 1
20	5 1 4 1 1 1 1
21	24 1 1 1
22	1 2 1 1 1 2
23	26 1 6
24	21 1 1 2
25	3 3 21 1 1 2
26	3 1 2 2 1 1 1 2
27	1 2 2 16
28	2 2 4 16
29	6 2 2 2 16
30	16 2 2 1
31	16 3 3 1
32	18 2 1 1 2 3
33	1 3 1 1 3 1
34	1 2 1 1 2 1
35	1 1 1 1 1 1 3
36	1 1 1 1 1 1 1 1 1 5
37	3 3 3 1 1 1 1 1 1 1 1 1
38	1 1 1 1 1 1 1 1 1 1 1 1 1 1 1
39	1 1 1 1 1 1 1 1 1 1 1 1 1 1 1
40	3 3 3 5 1 1 5 5
41	1 1 1 1 1 1 1 1 1 1 1 1
42	1 1 1 1 1 1 1 1 1
43	1 1 1 1 1 1 1 1 1
44	1 1 1 1 1 1 1 1 1
45	1 1 1 1 1 1 5
46	1 1 1 1 1 1
47	1 1 1 1 1 1
48	1 1 1 1 1 1
49	1 1 1 1 1 1
50	50

87

40x40 — 104 — shybobcat

Column clues (left to right, top to bottom):

1. 3 8
2. 2 2 4 3
3. 1 1 4 3 1
4. 1 2 2 4 11 1
5. 3 3 3 9 5 1
6. 1 12 12 9
7. 2 26 7
8. 28 1 7
9. 28 9
10. 30 7
11. 3 3 15 1 7
12. 1 2 2 13 6 2
13. 1 1 12 3 1
14. 2 2 7 4
15. 1 3 8
16. 1 1 9
17. 1 1 3 4
18. 1 1 2 5
19. 1 1 2 2 9 5
20. 1 1 2 2 22 1
21. 1 1 1 4 2 1 22
22. 1 1 1 8 22 1
23. 1 1 1 7 1 23
24. 1 10 24
25. 1 9 1 23
26. 1 2 5 20 1
27. 1 2 3 1 16
28. 1 3 15
29. 1 2 3 10
30. 2 2 9 5
31. 2 2 9
32. 4 2 3 4
33. 3 4 1 3 2 4
34. 3 4 3 2 4 9
35. 4 2 2 3 1 2 12
36. 2 3 2 2 2 2 2 14
37. 1 3 2 2 19
38. 1 3 22
39. 1 2 5 19
40. 26

Row clues (top to bottom, left to right):

1. 4
2. 14 3
3. 9 2
4. 1 5 9 2 4
5. 2 4 2 1 4
6. 1 3 9 2 4
7. 5 2 4 2 2 3
8. 5 5 4 2
9. 3 5 3 5 2 3 2
10. 2 9 2 5 2 2
11. 1 7 1 7 2 1
12. 2 9 2 2 6 2 3
13. 3 5 3 2 8 1 2
14. 5 2 9 3
15. 5 1 1 1 1 1 2 3 2
16. 7 2 2 1 1 1 2 2 2
17. 1 5 1 1 1 1 1 1 1 2
18. 2 6 1 12 2
19. 1 7 1 12 3
20. 1 7 1 12 1 1
21. 2 8 1 12 3 1 1
22. 1 8 1 12 2 5
23. 1 8 1 12 1 4
24. 2 10 2 12 2 5
25. 1 10 1 11 7
26. 1 10 1 11 2 4
27. 2 12 2 10 2 5
28. 1 12 1 9 2 5
29. 1 12 11 2 6
30. 1 12 11 2 6
31. 1 2 1 1 1 1 11 2 7
32. 1 12 9 2 7
33. 2 1 1 1 1 2 9 2 7
34. 15 7 2 7
35. 13 7 3 7
36. 11 7 10
37. 9 5 10
38. 7 5 9
39. 1 7 1 3 8
40. 1 9 1 5 6

88

50x45

104

eleninikolaoy

Column clues (left to right)

Column	Clue
1	13 1
2	13 2
3	12 3
4	12 3
5	11 3
6	11 3
7	10 2 5 4
8	10 3 7 5
9	9 6 12
10	9 7 3 4
11	9 8 11
12	8 9 9 2
13	8 8 10
14	7 10 12
15	7 26
16	7 24
17	7 24
18	7 24
19	7 22
20	7 21
21	7 20
22	8 18
23	8 16 1
24	9 4 11 1
25	9 3 11 2
26	9 1 11 3
27	9 10 4
28	10 10 4
29	10 13 5
30	11 14 5
31	12 18
32	13 17
33	13 19
34	14 20
35	13 20
36	12 6 12
37	12 6 10
38	12 7 9 1
39	11 7 8 2
40	11 7 2 5 3
41	10 6 5 4
42	10 6 15
43	9 6 12
44	9 6 10
45	10 6
46	10 6
47	10 6
48	11 6
49	11 6
50	11 5

Row clues (top to bottom)

Row	Clue
1	50
2	50
3	50
4	50
5	50
6	50
7	50
8	13 29
9	11 27
10	8 4 15 6
11	6 7 11 3
12	4 9 8
13	2 10 4
14	12 1
15	13
16	15 4
17	16 7
18	6 11 7
19	3 12 9
20	2 12 10
21	11 11
22	9 7 7
23	22 7
24	24 7
25	25 8
26	26 8
27	28 7
28	29 6
29	33 5
30	34 3
31	37
32	3 34
33	3 16 16
34	2 8 8 5
35	8 6 4
36	5 4 3
37	4 4 3
38	6 4 3
39	3 2 3 3
40	4 3 3
41	4 4 3
42	3 3 3
43	3 3 3
44	3 3 4
45	3 4 5

89

Column clues (top to bottom):

Column	Clues
1	50
2	1 5 1
3	1 5 1
4	1 5 1
5	1 3 1 1
6	1 1 2 1 2 1
7	1 3 2 2 2 1
8	1 2 1 2 3 1
9	1 3 2 3 3 1 1
10	1 2 1 3 3 2 1
11	1 3 1 6 2 1
12	1 3 2 6 5 1
13	1 3 1 5 4 2 1
14	1 2 1 4 5 4 1
15	1 3 1 3 3 3 1
16	1 3 1 2 3 5 1
17	1 3 1 4 2 1
18	1 4 1 2 2 1
19	1 5 1 3 2 2 1
20	1 8 3 1 3 1
21	1 19 7 4 1
22	1 3 1 1 1 7 4 1 6 1
23	1 7 3 1 1 1 5 2 5 2 1
24	1 1 7 1 3 18 1
25	1 1 6 1 1 2 7 1
26	3 6 1 1 1 7 1
27	1 2 3 1 2 13 1
28	1 2 1 1 10 5 1
29	1 2 3 5 2 1 3 1
30	1 2 8 1 1 6 1
31	1 2 6 1 3 9 1
32	1 2 6 1 6 2 1
33	1 2 5 2 2 18
34	1 7 1 3 3
35	1 4 3 2 2
36	1 3 3 1
37	1 4 4 1
38	1 11 1
39	1 4 1
40	50

Row clues (left to right):

Row	Clues
1	40
2	1 1 1
3	1 11 1
4	1 9 1
5	1 1 1 4 2 1
6	1 4 6 3 1
7	1 5 6 4 1
8	1 4 6 4 1
9	1 9 4 6 1
10	1 10 11 1
11	1 10 4 1
12	1 4 3 2
13	1 4 2 3
14	1 2 1 1 2 3
15	1 2 3 4 3
16	1 2 1 2 2 1
17	1 2 2 2 1
18	1 2 2 2 1
19	1 3 1 3 1
20	1 2 4 1 1 1
21	1 3 2 2 1 1
22	1 2 1 3 1 2 1
23	1 2 2 2 3 2 1
24	1 13 2 1
25	1 1 3 2 1
26	1 17 3 1
27	1 4 8 3 1
28	1 3 9 4 1
29	4 2 1 11 1
30	6 8 6 2 1
31	5 10 5 1 1
32	15 1 6 1 1
33	4 8 2 2 2 2 1
34	1 3 1 3 2 1 1
35	1 3 2 1 1 1 1 1
36	1 3 1 3 2 1 1 1
37	1 4 2 2 1 2 1 1 1
38	1 4 1 2 1 2 1 1 1
39	1 2 1 1 2 1 2 2 1 1
40	1 1 1 1 2 1 1 1
41	1 2 1 2 1 1 1 1
42	1 1 1 1 1 1 1 1
43	1 1 2 1 1 1 1 1
44	1 2 1 3 1 1 1
45	1 1 2 4 1 1 1
46	1 1 1 4 1 1 1
47	1 6 3 3 1
48	1 5 3 1
49	1 3 1
50	40

40x50

104

carm53

90

Column clues (top to bottom):

Column	Clues
1	11 2
2	24 2
3	2 2 2 2 2
4	2 1 2 10 2 2
5	6 2 1 2 1
6	8 1 2 1 1
7	2 2 2 8 1
8	2 3 1 2 9 1
9	4 7 15 2
10	3 6 2 2 4 8
11	2 1 6 2 3 6
12	1 1 4 3 2
13	3 15 2
14	7 2 1 2
15	1 1 3 2 2
16	2 2 3 2 2
17	1 2 4 2 10
18	1 2 3 1 1 2 4
19	1 1 2 1 1 15 2
20	2 3 2 1 7 2
21	16 2 4 2 1
22	6 4 2 5 4 2
23	5 1 6 4 6 1 1
24	2 2 1 5 11 2 1 1 1
25	1 2 2 1 2 7 2 4 1
26	2 2 1 3 3 8 5 1
27	1 2 2 4 7 3 1
28	2 2 2 1 8 11 1
29	1 2 2 20 1
30	2 2 10 2
31	1 2 1 2 2
32	2 2 2
33	1 3 1
34	2 2 1
35	1 2 1
36	2 3 1
37	1 2 2
38	1 3 1
39	1 2 1
40	1 2 1
41	4

Row clues (left to right):

Row	Clues
1	6
2	3 1
3	3 1
4	3 4
5	3 5
6	3 6
7	3 6
8	5 5
9	8 4
10	4 3 4
11	5 6
12	3 2 4
13	2 1 2 2
14	1 1 3 1
15	2 5 1
16	5 1
17	1 1 1
18	5 1 5
19	8 1 2 3
20	4 3 1 4 2
21	2 2 2 1 1 6 2
22	2 2 2 1 1 2 2 1 2
23	2 3 6 4 2 2 1
24	2 2 5 4 2 2 2
25	3 2 2 4 2 2 1
26	4 3 7 3 3 2
27	2 2 3 9 3 3
28	2 3 2 2 3 3
29	2 3 19
30	3 2 2 3 10
31	4 1 1 10
32	2 1 3 2 6
33	1 1 2 2 6
34	1 1 2 2 6
35	1 1 2 1 6
36	1 1 1 1 2 1 1
37	1 1 1 1 2 2
38	1 1 1 1 1 2
39	1 1 1 2 1 2
40	1 3 2 2 1 2
41	1 3 1 2 2 2
42	1 2 1 1 2 1 2 2
43	1 2 1 1 4 3 2
44	2 2 1 1 4 1 2
45	2 2 2 1 2 5
46	2 2 2 1 1 6
47	7 3 4
48	5 2 4
49	4 13 8
50	8 10 10 4

41x50

104

TNT

91

50x40 · 104 · griddlock

Column clues (left to right, top to bottom):

1. 20
2. 22
3. 4 15
4. 3 16
5. 3 18
6. 3 18
7. 2 18
8. 3 18
9. 4 18
10. 5 18
11. 5 18
12. 4 18
13. 4 18
14. 2 18
15. 6 18
16. 7 18
17. 8 3 9
18. 2 2 7
19. 2 2 14
20. 2 2 6 7
21. 2 2 2 3 4
22. 2 3 2 3 4 3
23. 2 3 1 3 4 4 3
24. 2 3 1 2 3 7 3
25. 2 3 2 2 2 8 2
26. 2 3 1 2 2 10 2
27. 2 3 1 2 1 10 3
28. 2 3 3 2 12 2
29. 2 3 1 12 2
30. 2 3 2 13 2
31. 2 2 1 13 2
32. 2 2 1 14 2
33. 8 2 1 14 2
34. 7 2 1 14 2
35. 2 2 1 13 2
36. 2 2 2 13 2
37. 2 2 1 13 2
38. 2 2 1 12 2
39. 2 3 2 11 2
40. 2 2 2 10 2
41. 2 3 2 8 2
42. 2 3 2 7 2
43. 2 3 3 5 2
44. 2 3 8 2
45. 3 3 3
46. 3 6 4
47. 3 16
48. 5 14
49. 22
50. 18

Row clues (top to bottom, left to right):

1. 17
2. 19
3. 3 2
4. 2 3 2
5. 5 3 2
6. 6 3 8 2
7. 13 8 14
8. 14 7 15
9. 4 4
10. 3 3
11. 3 11 2
12. 3 4 12 3
13. 2 2 5 6 3
14. 2 2 3 4 2
15. 2 2 3 7 3 2
16. 2 2 3 3 4 3 2
17. 2 14 2 3 2 4 2
18. 17 2 2 2 4 2
19. 17 2 2 2 6
20. 16 2 1 2 5
21. 16 2 2 1 5
22. 16 2 1 2 5
23. 16 2 1 3 1 4
24. 16 2 2 11 1 4
25. 16 1 1 15 1 4
26. 17 1 1 19 1 4
27. 17 1 23 4
28. 20 22 4
29. 20 22 4
30. 20 21 4
31. 20 20 4
32. 19 19 4
33. 19 17 3
34. 19 15 2
35. 3 12 2
36. 3 8 2
37. 4 3
38. 4 5
39. 13
40. 7

92

50x40

104

xiayu

93

														1												2	2	2											
													1	2	1	1										2	1	1											
								1	1	1	1	1	3	1	2	2	1								2	2	2	8					1	1					
						1	1	2	5	6	2	3	2	2	1	1	1	1							1	2	2	1		1		1	2	1					
				1	1	2	2	3	1	1	3	2	2	2	2	1	1	1	3	2	2	2			2	3	2	2	3	5	1	2	7	7	1				
			1	4	3	2	5	2	2	2	2	2	2	2	2	2	1	1	1	1	1	1	2	2	4	4	2	4	2	1	4	2	2	4	6				
	1	1	8	2	2	1	2	2	2	2	2	2	2	1	2	2	2	2	2	2	3	6	6	8	1	1	4	2	3	3	1	2	5	1	4	1	1		
42	4	8	2	3	5	5	5	3	2	3	2	2	2	2	1	3	2	6	4	1	1	2	3	1	2	1	1	1	5	2	2	5	4	2	3	2	3	1	35
4	4	3	2	1	1	1	1	1	2	3	8	7	1	1	3	1	5	4	6	4	3	1	5	3	2	3	3	1	5	8	10	2	2	2	2	2	1	12	10

							40
					1	11	1
				1	5	2	1
				1	4	1	1
				1	3	1	1
				1	3	5	1
			1	4	3	3	1
			1	2	2	1	1
			1	2	2	3	1
			1	2	2	7	1
		1	3	2	1	1	1
		1	3	2	2	1	1
		1	3	2	1	1	1
		1	10	2	1	1	1
				1	4	2	1
				1	3	2	1
				1	1	2	1
				1	2	2	1
				1	3	2	1
				1	2	2	1
				1	1	2	1
				1	2	2	1
			1	5	2	3	1
			1	13	2	2	1
			1	4	3	4	1
			1	2	2	1	1
		1	2	7	1	10	1
		1	1	9	2	10	1
1	2	2	2	1	1	2	1
1	2	2	2	1	3	3	1
1	1	2	2	2	2	1	1
1	1	2	2	2	2	1	1
		1	1	14	7	1	1
			1	2	12	4	1
				1	2	3	1
				1	2	2	1
				1	1	2	2
				1	1	2	2
				1	7	2	2
			1	4	4	3	2
				3	2	2	2
				2	2	3	2
			1	3	1	2	2
			2	5	2	2	2
	1	5	6	4	3	2	2
2	5	3	4	2	1	4	2
2	3	2	8	3	1	3	2
		2	3	2	5	2	13
		3	4	1	5	5	8
	9	3	2	2	2	2	2

40x50

104

oko

94

Column clues:

												3	4	4	5	3	3																																
											4	1	1	1	1	1	1																	1	2		2												
										3	3	1	1	1	2	1	1	2					2				1	1				1		1	1	3	1	2											
								6	7	3	2	2	1	2	1	2	2	1	2			2	1	3			1	1	1	1	1	1	2	1	1	1	1	1											
							5	2	2	2	1	1	2	2	5	2	4	1	1	2	2	1	1	2			3	2	2	2	1	1	1	1	1	2	2	1	3	4									
					6	5	3	2	1	1	5	4	2	2	2	3	3	1	1	1	2	2	1	1	1		2	1	2	2	2	8	1	2	2	2	1	2	1	1									
					1	4	6	2	3	4	1	1	2	1	2	1	2	1	1	1	2	1	3	4	1	1	1	3	1	1	1	2	1	5	3	2	3	1	1	2	5								
				7	2	3	1	2	2	1	1	1	2	2	3	4	1	4	1	1	1	1	2	1	1	1	2	2	2	4	2	1	3	2	3	1	2	5	2	1	2	3							
			6	7	2	2	2	3	1	1	3	6	7	1	2	2	3	2	4	2	2	2	1	2	8	1	1	1	1	2	2	2	2	2	1	2	1	1	1	3	2	3	3	3	4				
		5	4	4	2	2	3	2	4	7	2	2	3	2	1	1	2	2	3	1	1	1	1	1	1	1	1	1	2	1	1	2	4	3	3	1	3	1	7	1	2	6	3	4	1				
		7	1	1	8	3	3	7	1	6	1	2	1	2	3	2	4	2	2	2	2	8	2	2	2	3	3	2	7	3	2	3	1	2	1	1	2	4	1	1	1	2	2	2	7	4	2	4	
11	15	4	10	7	3	4	3	1	1	3	4	1	2	2	3	4	4	5	6	13	12	1	1	1	1	1	1	1	3	2	2	2	2	5	3	2	2	1	1	3	4	3	2	1	1	1	2	4	4	8

Row clues:

19
12 3
7 2 2
7 4
14 12
5 3 4 3
4 3
17 14
8 2 2 3 4
12 3 3 6 3
4 3 2 3 2
3 4 1 2
4 2
4 7 8 1
14 15
4 2 4 3 7 3 2
3 2 6 6 1 7 2 2
2 2 2 2 1 1 3 2 2 2
2 1 1 3 2 2 1 1 1 2 2 1 1 1
2 1 1 4 1 1 2 2 1 1 4 1 1 1
2 1 1 5 1 1 2 1 1 1 4 1 1 1
3 1 1 4 1 2 4 3 1 2 4 1 1 1
3 1 2 2 2 1 2 2 2 1 2 2 1 2
3 2 2 2 2 2 1 1 3 3 2 2
3 2 5 1 1 2 2 2 2 3 2 1
4 1 2 5 6 3 2 2
4 3 3 2 2 7 1
4 6 2 2 1
3 4 3 2 2
5 3 1 1 6
5 2 2 1 3 3
3 2 2 1 3 1 2
5 3 1 2 4
4 5 5 1 2 3
1 2 4 5 2 5 2
1 2 3 4 14 2
2 1 2 2 3 4
1 1 1 1 5 3
1 1 2 1 4 4
1 1 1 1 4 10
4 1 1 2 1 3 1
2 1 1 1 1 1 3 1
2 3 1 2 2 3 2
2 2 3 2 3 3
2 2 1 4 5
3 2 11 2
3 4 12 2
3 3 7 3 2
2 2 9 9
33

50x50

104

xiayu

95

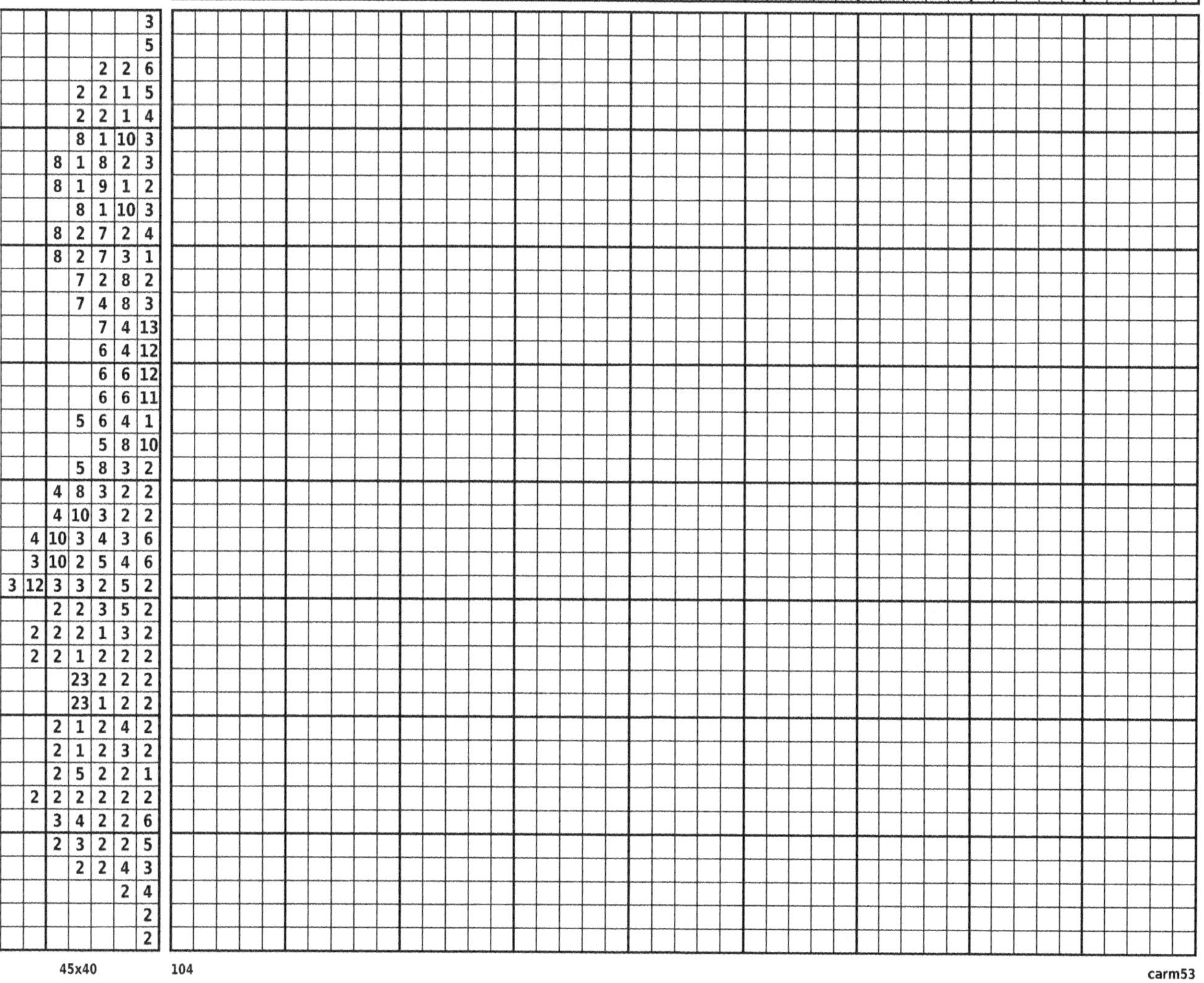

																						4																						
																						2	2	4	5	7																		
																						6	2	2	1	2	9																	
														4	7	9						1	2	2	4	4	8																	
		20				12	9	6	3	3		3	3	10	7	4	12	14			20	3	5	4	1	1	2	7																
	20	2	20	18	15	1	4	7	10	13	20	16	13	2	2	2	1	5	18	20	4	2	1	1	2	3	2	5	5	2											2	2		
2	2	3	9	10	5	2	2	2	2	2	2	2	2	2	2	4	7	1	8	9	7	3	1	4	2	3	2	19	18	2	2	3	3	3	3		2	2	16	17	3	3	3	3

						3
						5
				2	2	6
			2	2	1	5
			2	2	1	4
			8	1	10	3
		8	1	8	2	3
		8	1	9	1	2
			8	1	10	3
		8	2	7	2	4
		8	2	7	3	1
			7	2	8	2
			7	4	8	3
				7	4	13
				6	4	12
				6	6	12
				6	6	11
			5	6	4	1
				5	8	10
			5	8	3	2
		4	8	3	2	2
		4	10	3	2	2
	4	10	3	4	3	6
	3	10	2	5	4	6
3	12	3	3	2	5	2
		2	2	3	5	2
	2	2	2	1	3	2
	2	2	1	2	2	2
			23	2	2	2
			23	1	2	2
		2	1	2	4	2
		2	1	2	3	2
		2	5	2	2	1
	2	2	2	2	2	2
		3	4	2	2	6
		2	3	2	2	5
			2	2	4	3
					2	4
						2
						2

45x40

104

carm53

96

Column clues (left to right, top to bottom):

Column	Clues
1	6 4
2	14 6
3	2 1 12 3
4	2 1 2 2 4
5	3 8 2 2 3
6	1 3 1 2 1 1 3
7	2 1 1 1 1 1 2 2
8	2 2 1 1 1 1 2 1 2
9	1 9 5 1 1 2 1 2
10	2 8 12 1 2
11	5 1 1 2
12	2 7 1 2
13	1 9 1 1 1 2
14	3 2 1 1 1 1 2
15	1 2 2 1 1 1 1 2
16	1 1 2 1 1 1 1 2
17	4 4 1 1 1 1 2
18	1 5 1 1 1 1 2
19	2 2 3 1 1 1 1 2
20	9 2 2 1 1 1 2
21	2 8 4 1 1 1 2
22	1 2 1 1 1 1 2
23	1 3 2 1 1 1 2
24	5 3 1 1 1 1 2
25	1 1 2 1 1 1 1 2
26	2 2 1 1 1 1 1 2
27	3 2 1 1 1 1 2
28	1 2 1 1 1 1 2
29	1 6 1 1 1 1 2
30	1 4 1 1 1 2
31	2 8 2 6 1 1 2
32	9 2 1 2
33	2 3 1 5 1 2
34	1 10 1 1 2
35	1 3 2 1 1 1 2
36	5 2 1 1 1 2 2
37	1 1 1 1 1 1 1 2
38	2 1 2 2 1 1 1 2
39	4 5 1 2 1 2
40	1 2 1 1 3
41	2 2 1 1 2 1 4
42	1 9 2 1 2 3
43	2 8 14 2
44	2 4 8
45	4 5

Row clues (top to bottom):

Row	Clues
1	1 1 1 1 1
2	1 1 1 1 1 1
3	1 1 1 1 1
4	1 3 2 1 1
5	3 3 3 1 1
6	4 2 3 1
7	1 2 2 3 3 3
8	2 3 2 1 1 2 3
9	2 1 1 2 1 1 2 3 2
10	2 2 2 1 2 1 2 2 1 2 2
11	4 2 22 1 1 2
12	1 1 4 1 3 1 2
13	1 4 1 1 5 2
14	3 1 1 4
15	2 1 2 1 2 1 2
16	2 3 1 4 5 1 2 1
17	2 1 1 2 1 2 1 2 1 1
18	2 2 1 1 1 1 1 1 1 2
19	2 1 1 1 2 2 1 2 1 1
20	2 1 1 1 1 1 1 1 1 1
21	1 1 2 1 1 2 2 1 2 2
22	2 2 1 1 1 2 1 6 1
23	2 1 4 1 1 1 1 2 1 2
24	2 1 1 1 3 8 1 4
25	1 2 1 6 4 1 1 2 1
26	2 1 1 1 1 5 1
27	1 4 1 1 1 1
28	1 1 20 2 3
29	3 1 1 1 1 3 1
30	8 1 1 1 3 1
31	1 3 1 5 2
32	3 24 3
33	5 4
34	2 3 10
35	2 3 2 2
36	4 25 2
37	3 4
38	4 5
39	39
40	37

45x40

105

xiayu

97

Column clues (top to bottom, columns left to right):

Column	Clues
1	35
2	3 1 2 2
3	2 1 1 2 1
4	2 1 1 2 1
5	1 1 1 1 1 2
6	2 1 1 1 1 2
7	2 1 1 2 1 2
8	2 1 1 1 1 1
9	2 1 6 4 2 18 1
10	1 4 1 5 2 1 1 1
11	1 3 1 3 2 1 1 1
12	1 2 1 1 4 2 1 3 1 1
13	2 2 1 11 1 6 1 1
14	2 1 1 4 2 2 8 1 1
15	1 2 1 5 1 2 4 5 1 1
16	1 1 1 6 1 1 1 6 1 1
17	1 1 1 3 3 2 1 1 3 1 1
18	1 1 1 3 3 1 2 16 1
19	1 7 2 8 1 6 1 1
20	1 1 1 5 1 11 1 1
21	2 1 1 1 9 2 1 1
22	2 1 1 1 1 5 2 1 1
23	3 5 2 1 4 2 1 1
24	1 2 1 4 1 1 1 2 1 1
25	2 1 1 2 29 1
26	1 3 1 2 1 1
27	1 1 1 2 2 1
28	2 2 3 1 1 1 1 2
29	2 1 3 1 2 1 1
30	3 12 1 1 1
31	2 1 20 1
32	3 1 1 1 1
33	3 1 1 1 1
34	2 1 1 4
35	34

Row clues (left to right, rows top to bottom):

Row	Clues
1	6
2	2 2
3	17
4	2 1 1 1 3
5	2 2 1 2 2
6	2 2 1 1 2
7	2 2 1 1 1 2
8	1 1 1 1 1 2
9	5 2 1 1 7
10	2 3 7 7 1
11	1 1 1 1 2 2
12	2 1 1 1 1 1
13	1 1 2 1 2 2
14	1 1 4 1 1 2
15	1 1 6 1 2 1
16	1 1 2 4 1 1 1
17	1 1 7 1 1 1 1
18	2 1 11 1 1
19	1 6 11 11
20	1 1 2 1 5 1 1
21	1 1 2 3 1 1 1
22	1 1 3 1 1 1 1
23	1 1 7 3 1 1 1
24	1 4 6 1 1 1
25	1 3 1 4 1 1 1
26	1 2 3 4 1 1 1
27	1 1 6 6 1 1
28	1 1 1 4 1 1 1 1
29	1 1 2 2 1 1 1
30	2 4 1 2 1 1 1
31	7 12 1 1 1 1
32	1 3 2 10 7
33	1 21 1 1
34	1 1 6 1 1 1
35	1 1 2 3 1 1 1
36	1 1 2 1 1 1 1 1
37	1 1 1 2 1 1 1 1
38	1 1 2 2 1 1 1 1
39	1 1 1 1 1 1 1 1
40	1 1 1 1 1 1 1 1
41	1 1 1 1 1 1 1
42	1 1 1 1 1 1
43	1 1 1 1 1 1
44	1 1 1 1 1 1
45	1 2 1 1 1 1 2
46	10 1 1 9
47	1 7 9 2
48	4 6
49	3 1
50	22

35x50

105

xiayu

98

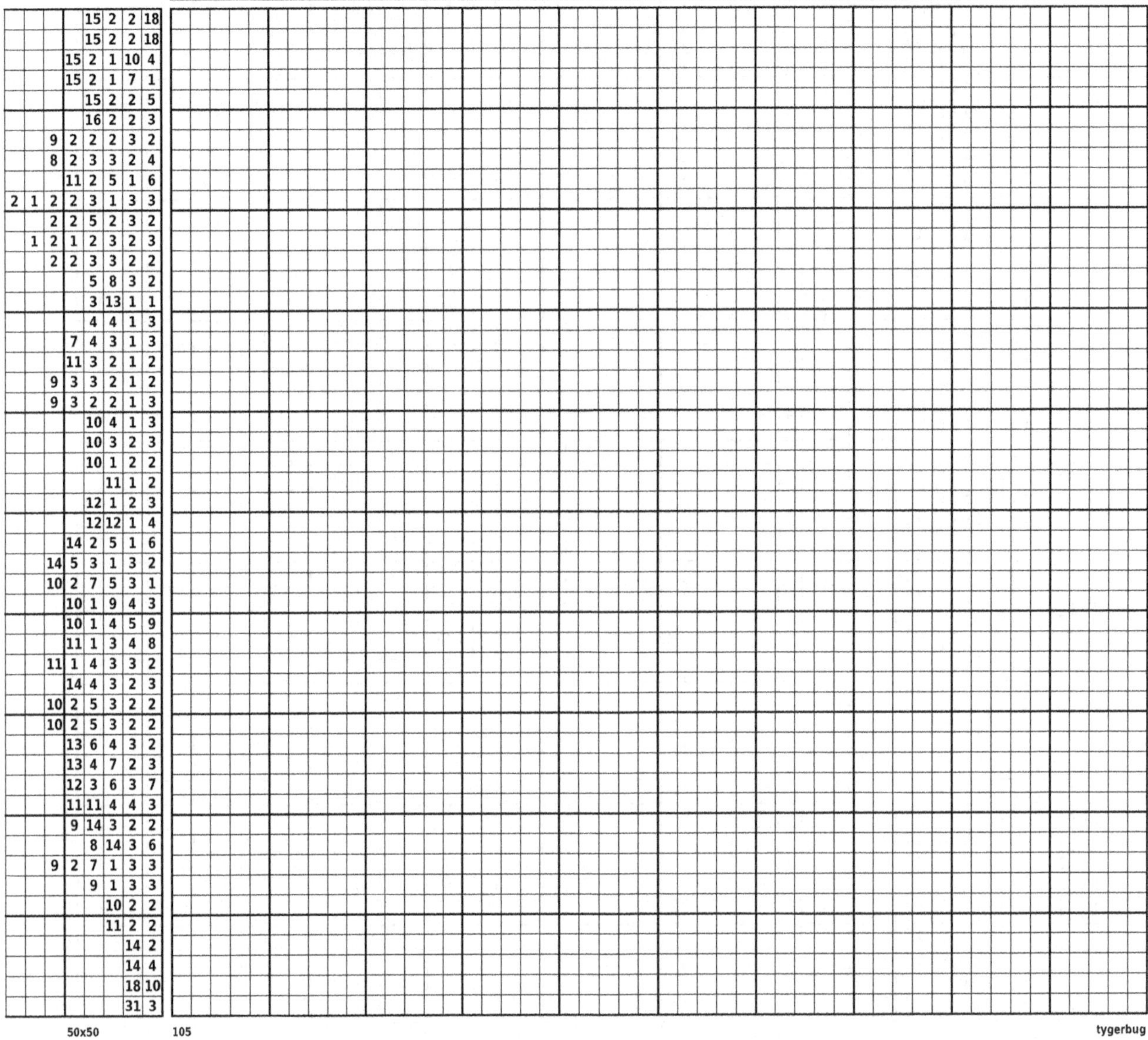

99

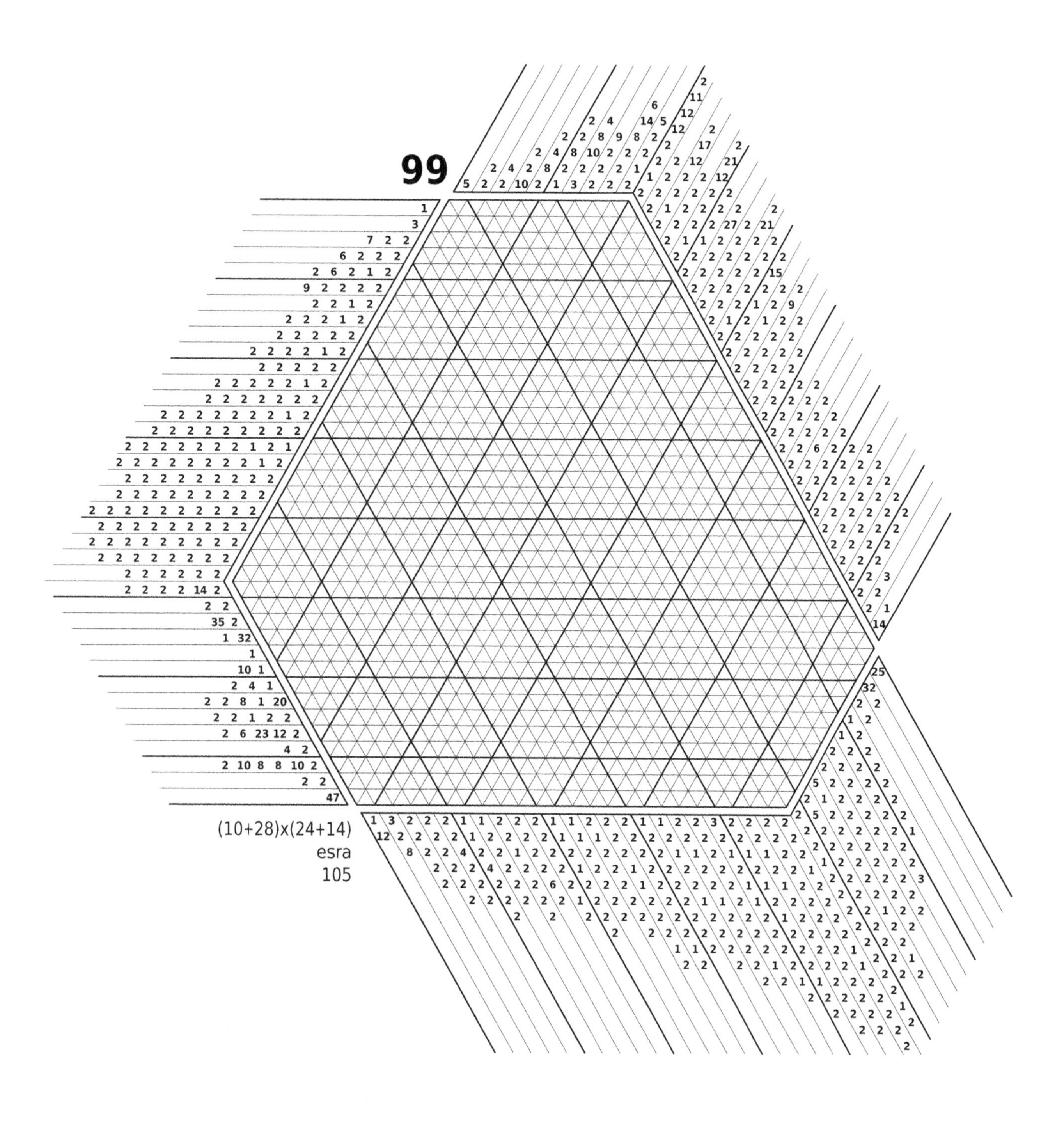

(10+28)x(24+14)
esra
105

100

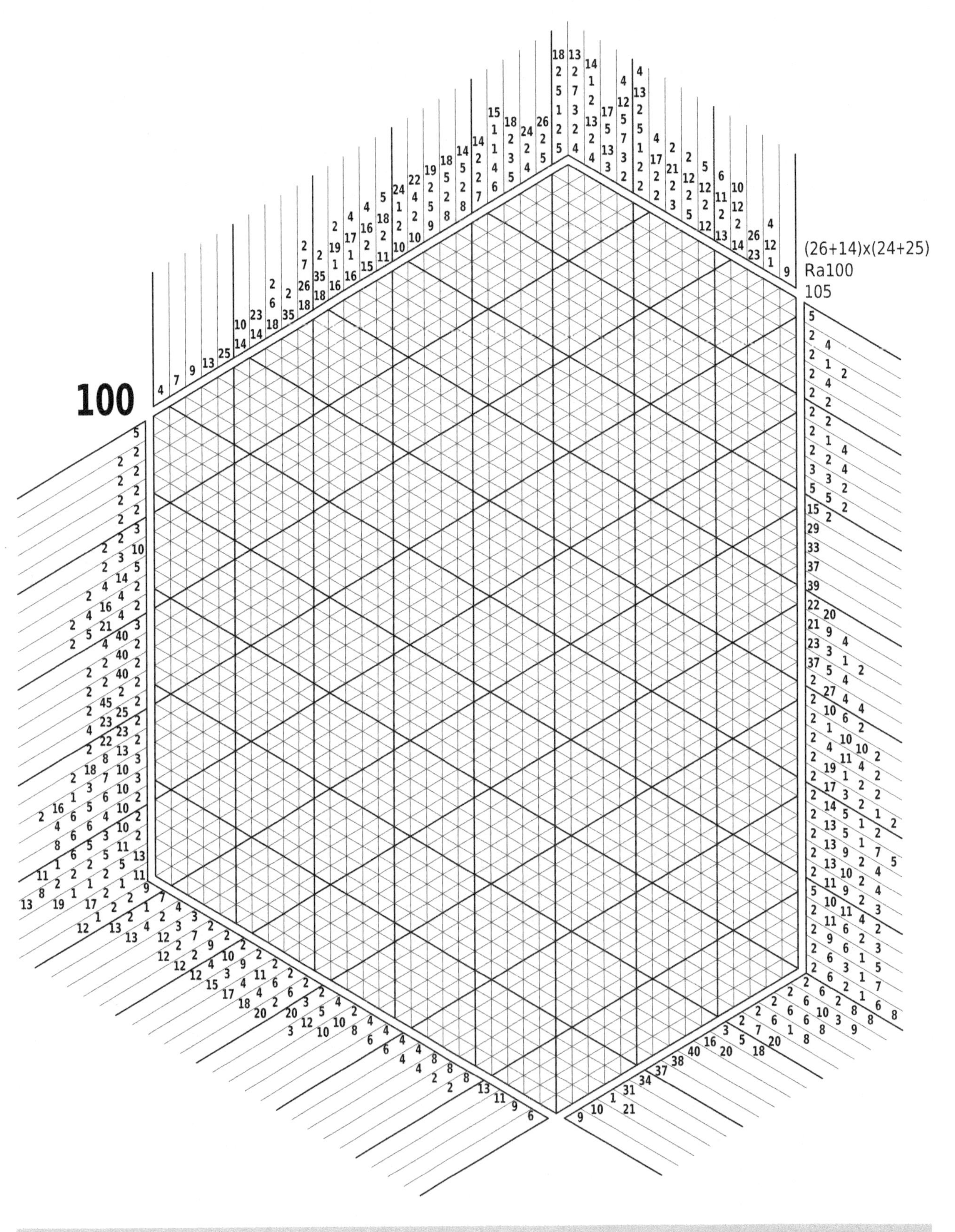

101

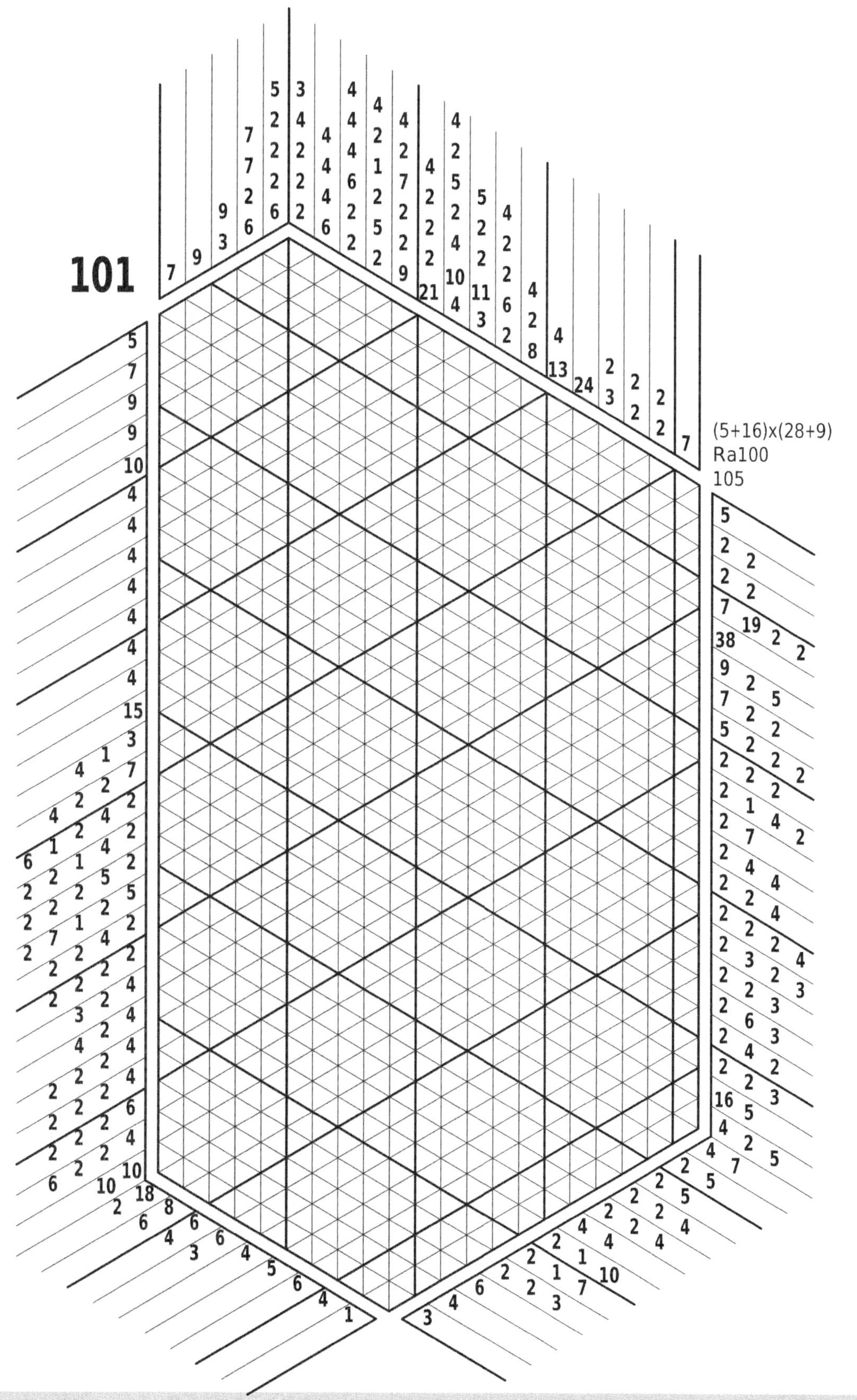

(5+16)x(28+9)
Ra100
105

102

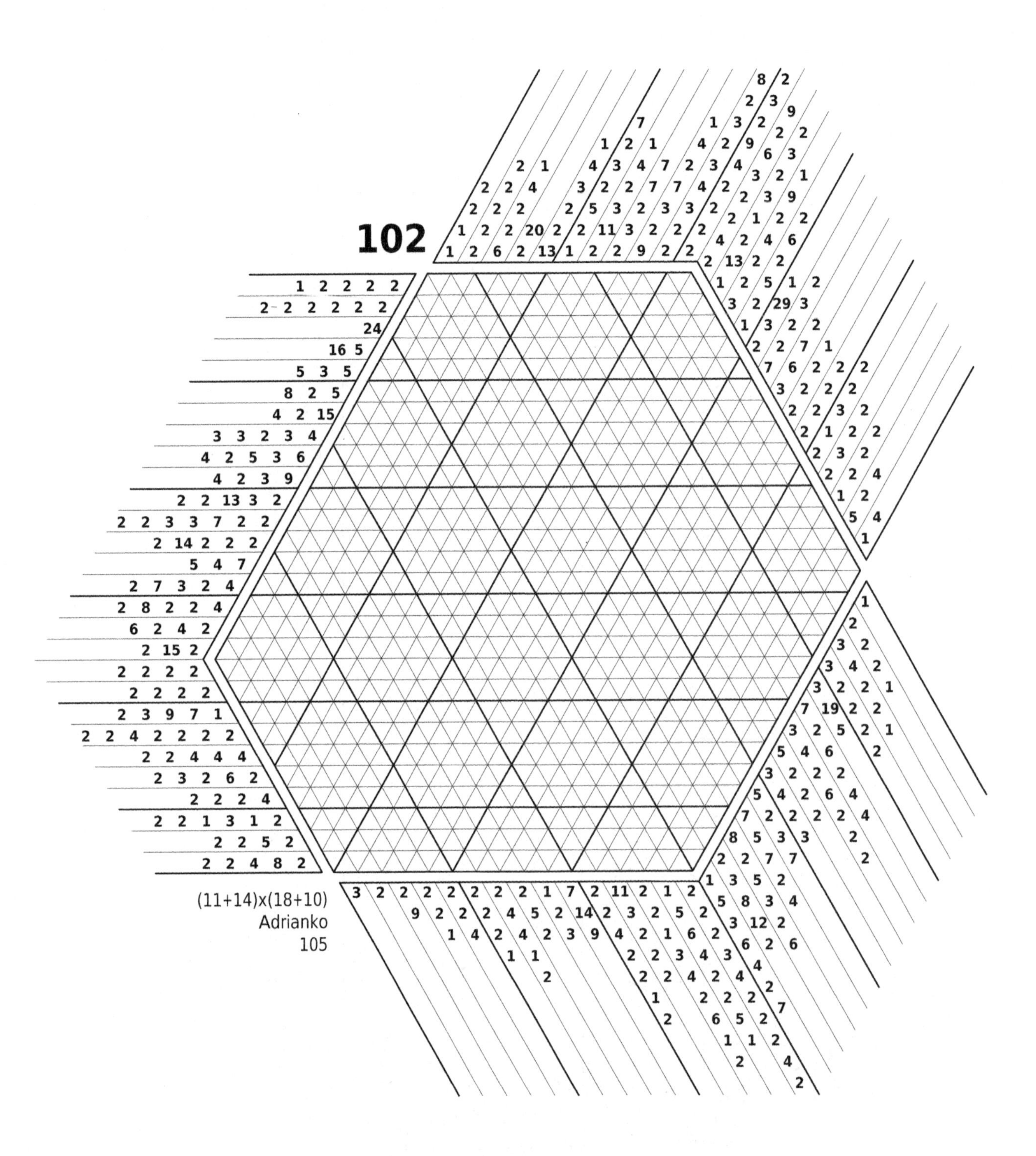

(11+14)x(18+10)
Adrianko
105

103

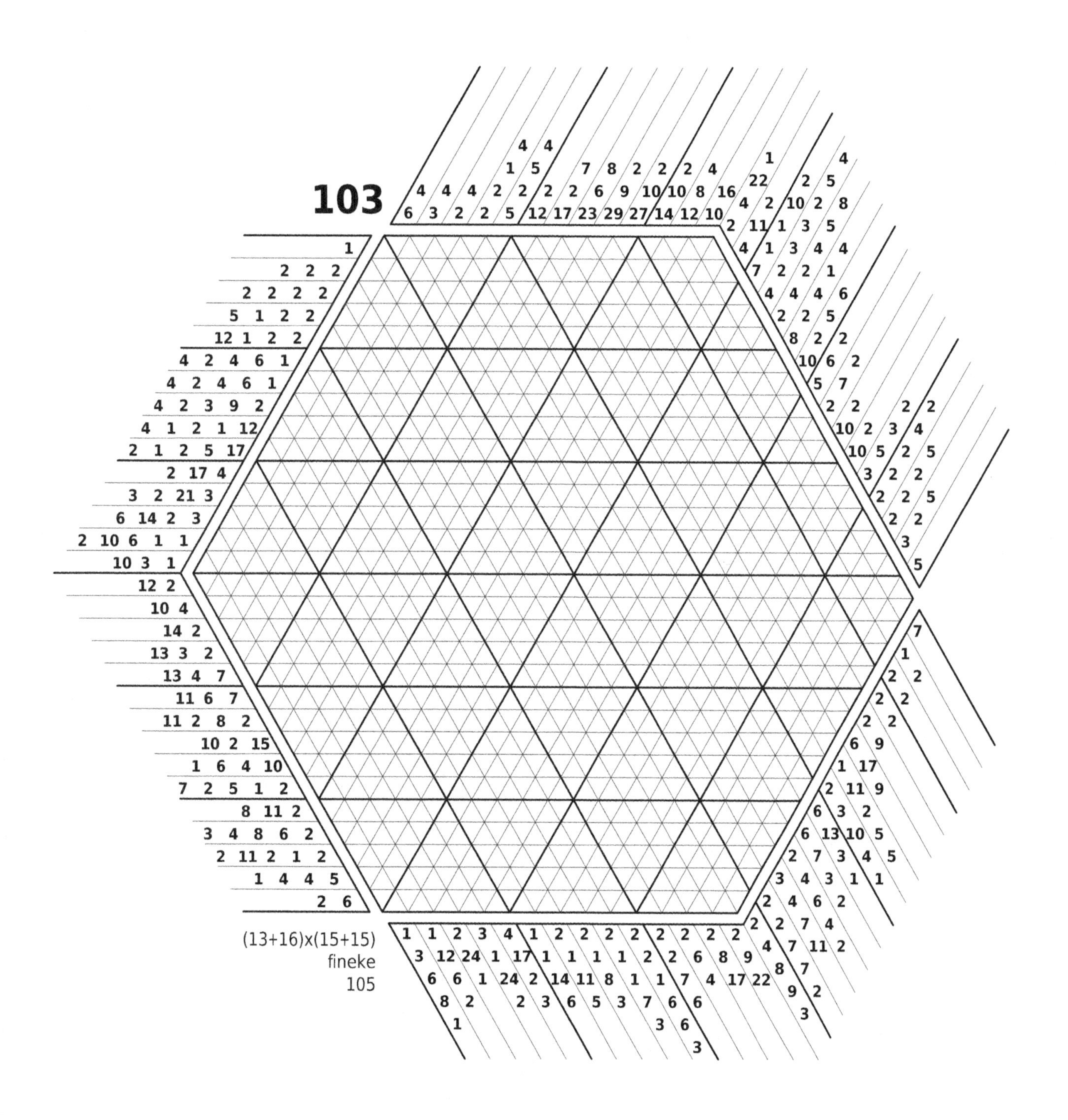

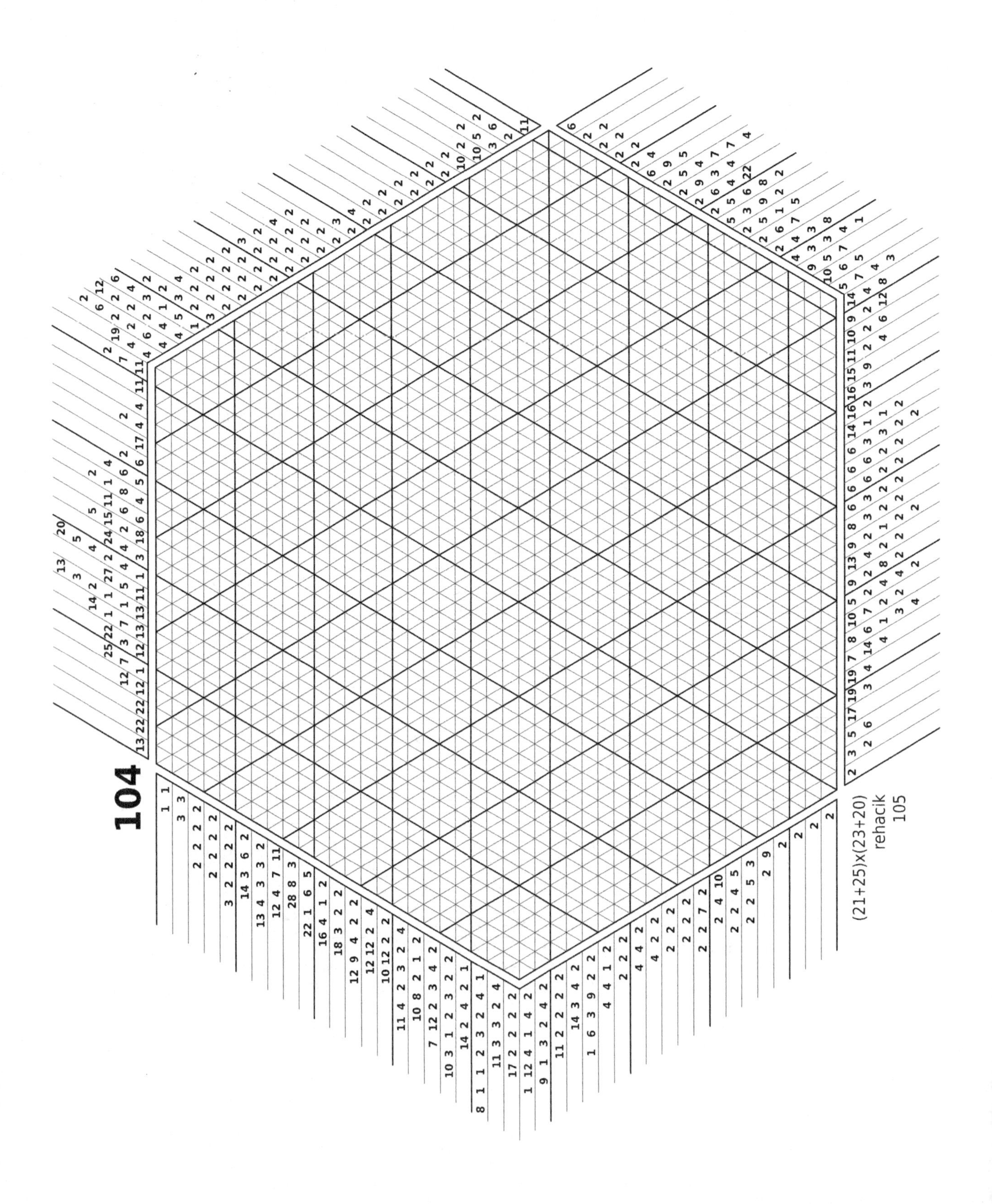

Solutions

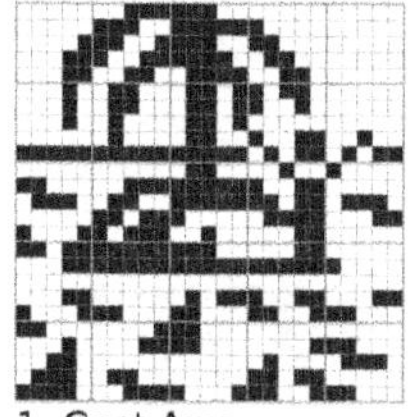
1: Cast Away

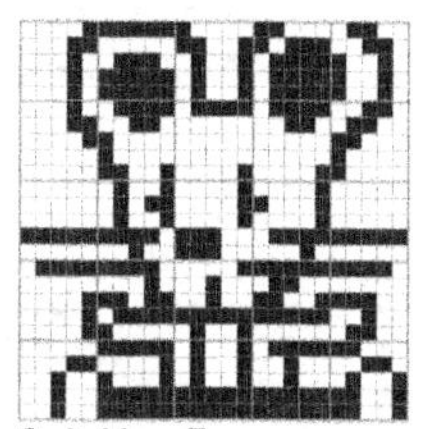
2: It Has Fur...

3: New York

4: Smoker

5: Zebra

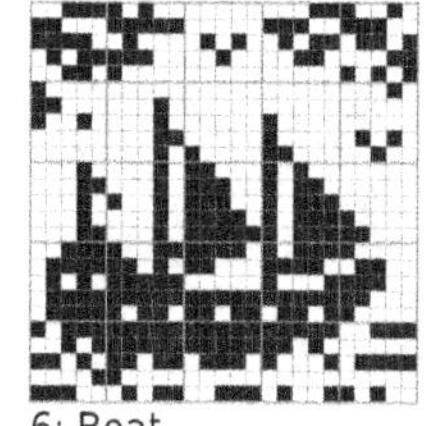
6: Boat

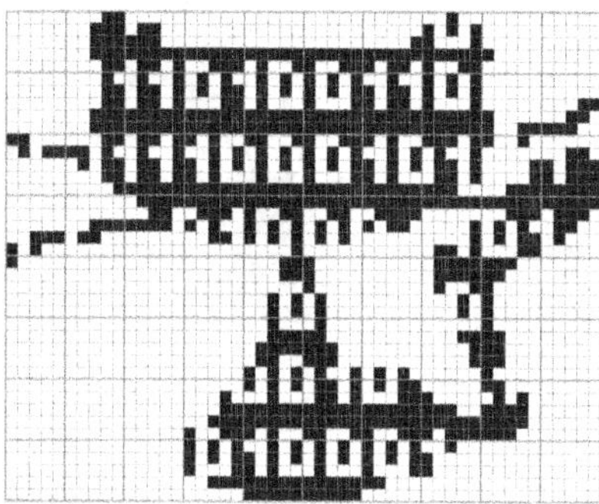
7: Binary Cat

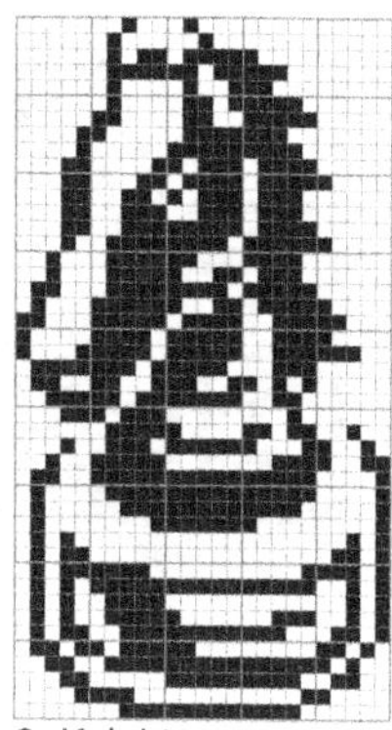
8: Knight

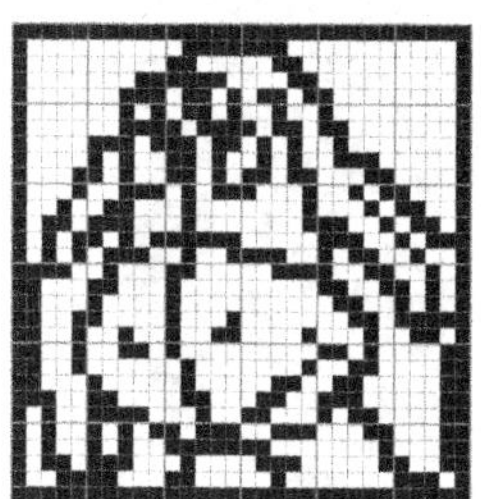
9: Eeyore

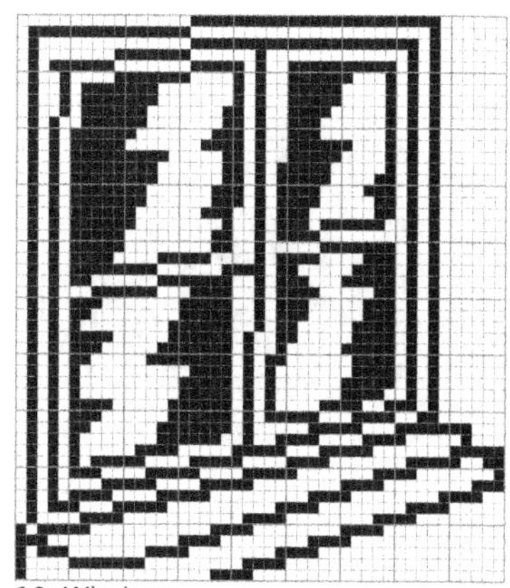
10: Windows

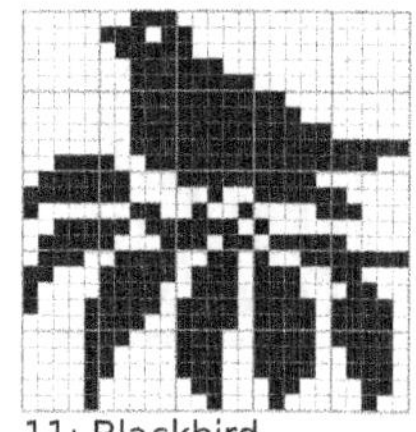
11: Blackbird

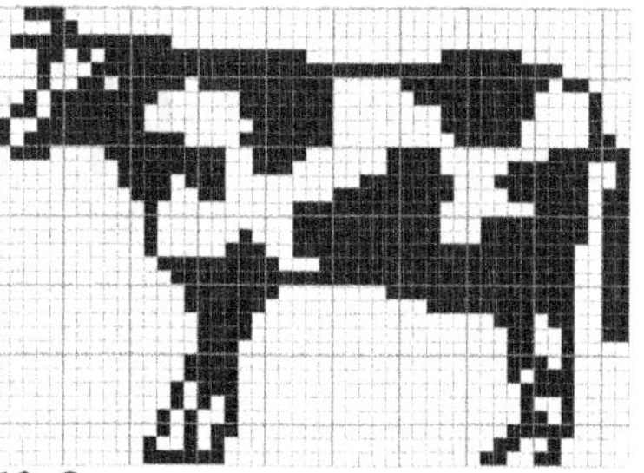
12: Cow

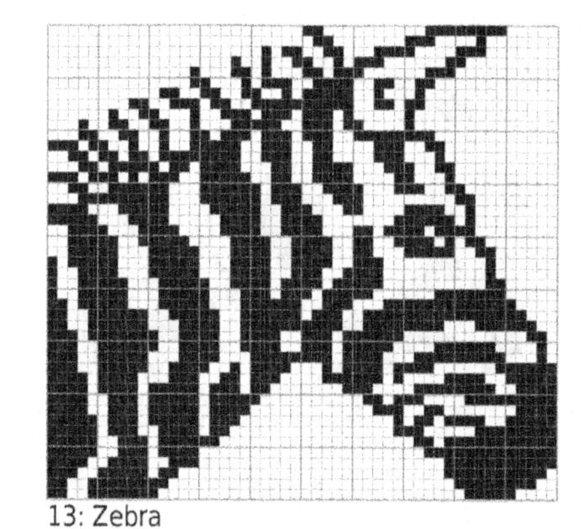
13: Zebra

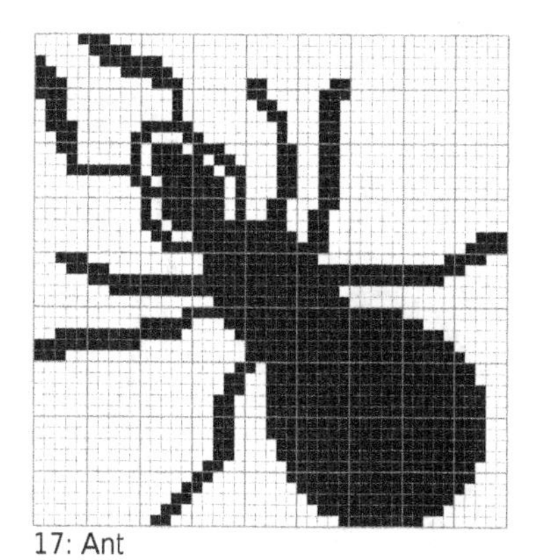
17: Ant

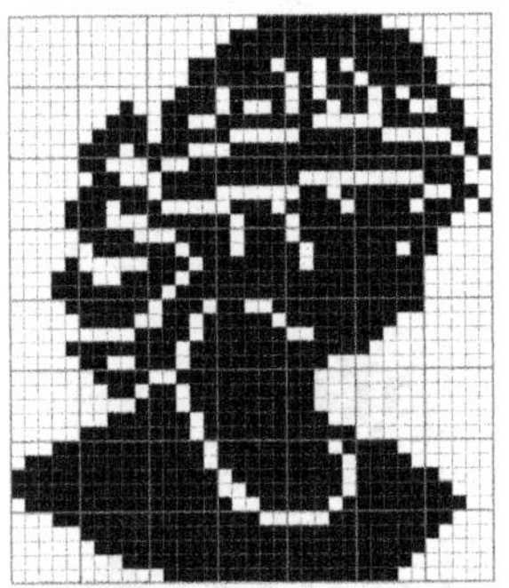
18: Silhouette2

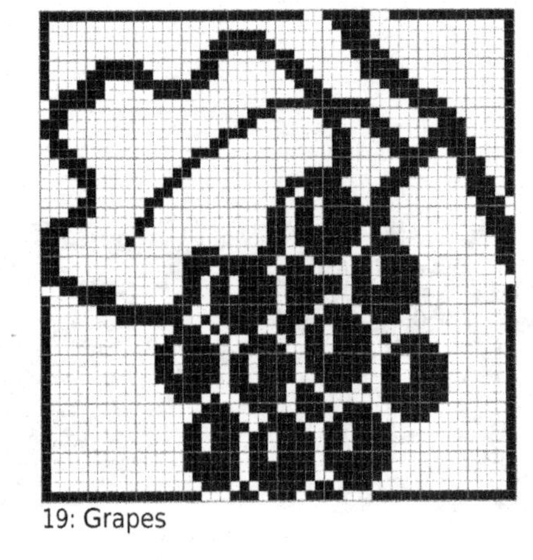
19: Grapes

21: Tank

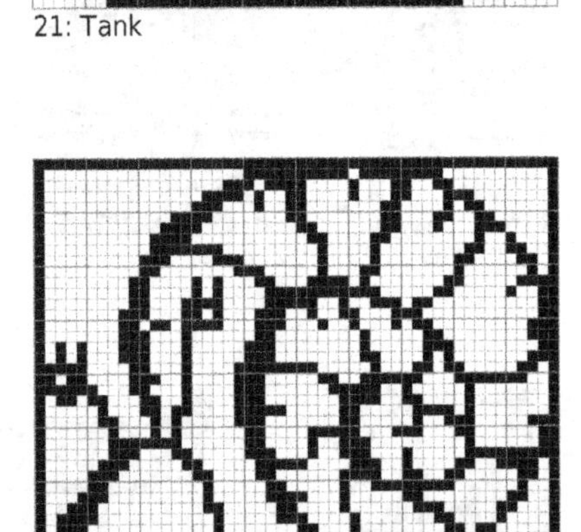
23: Snail

22: Clown

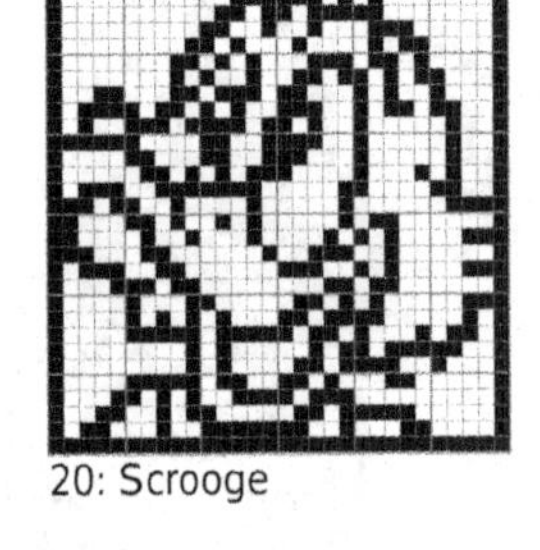
20: Scrooge

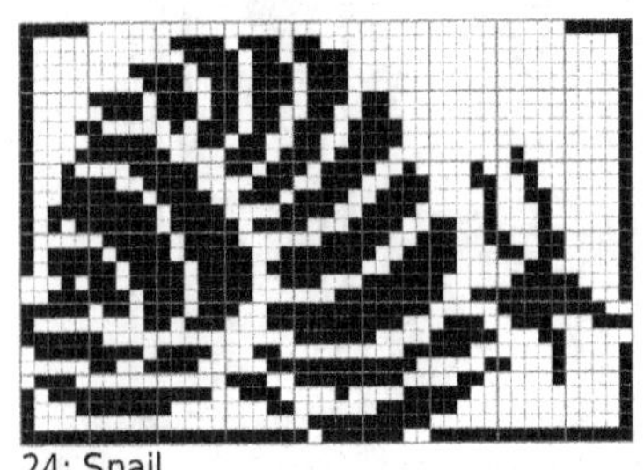
24: Snail

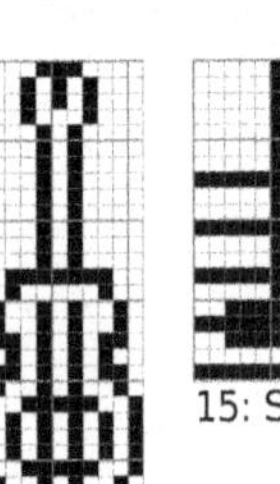
14: Guitar

15: Sol

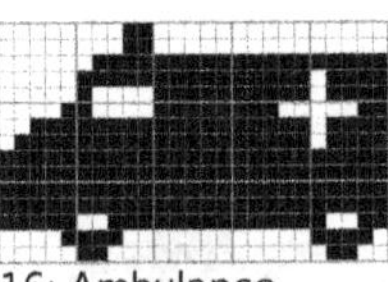
16: Ambulance

25: Happy Birthday!

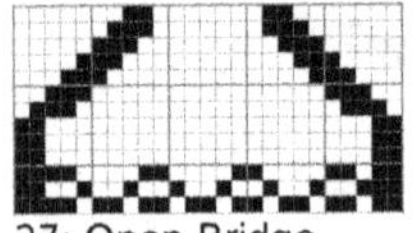
27: Open Bridge

26: Dinner

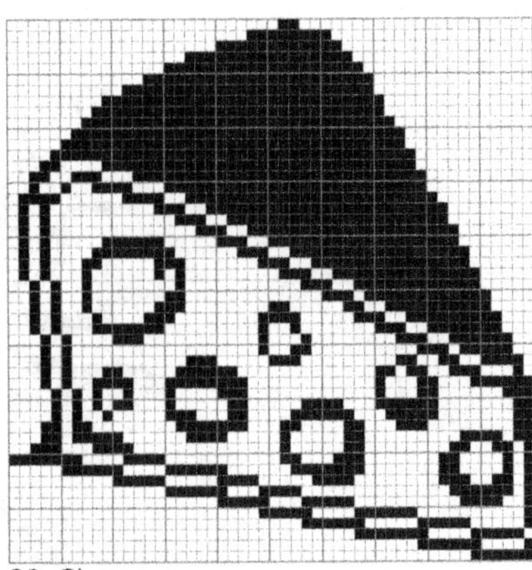
28: Cheese

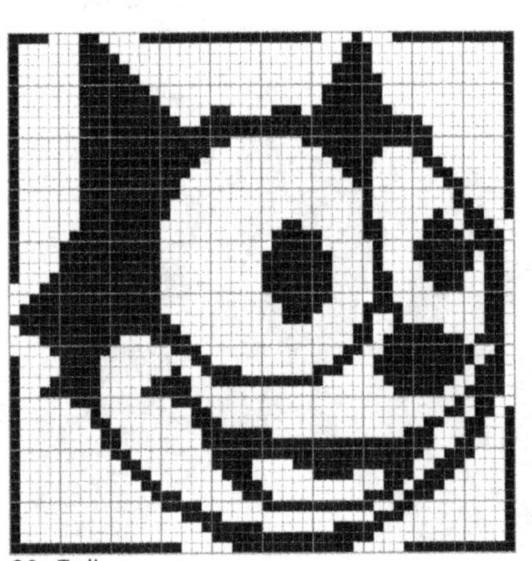
29: Felix

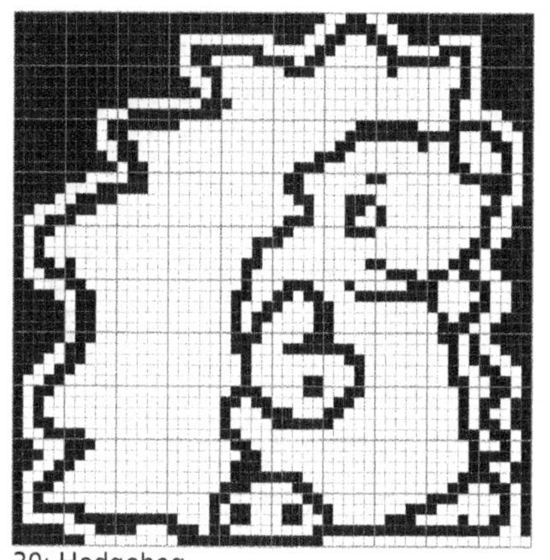
30: Hedgehog

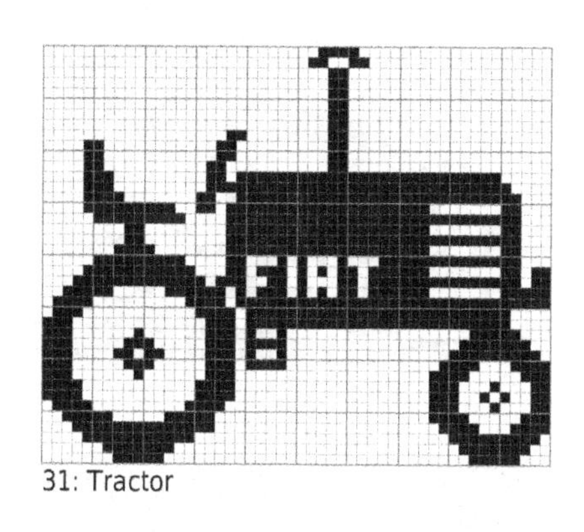

31: Tractor

32: Cornus Florida

36: Lonely Island

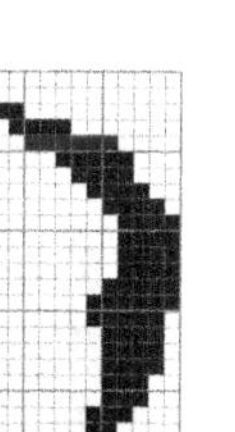
34: Scarlet Rose

35: Gramophone

33: Devil

37: Cow

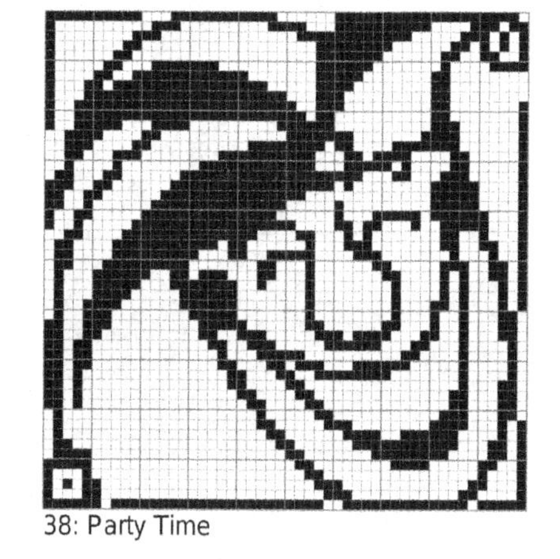
38: Party Time

39: Cupid

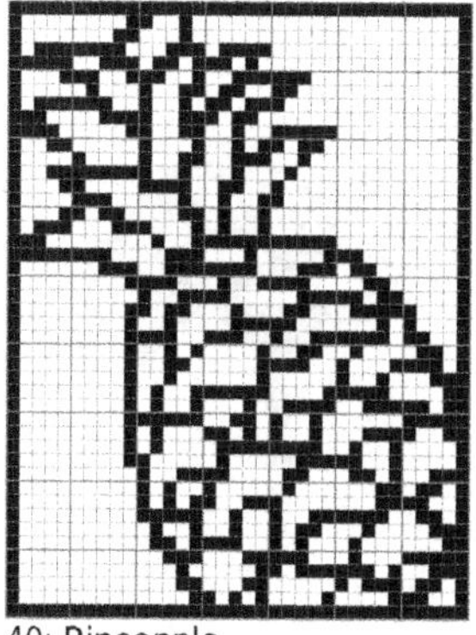
40: Pineapple

41: Mario

42: Birds

43: Lizard

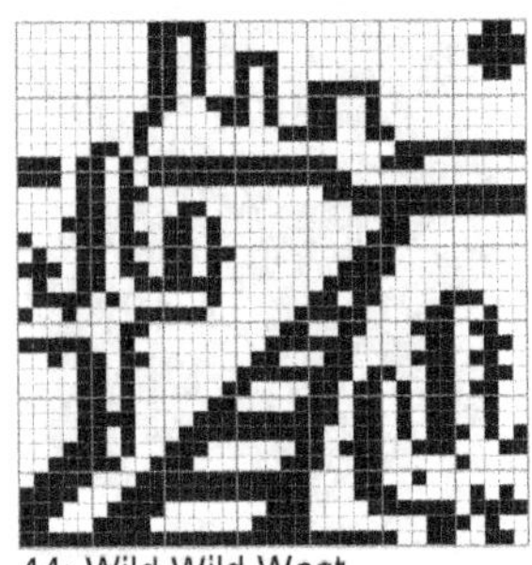
44: Wild Wild West

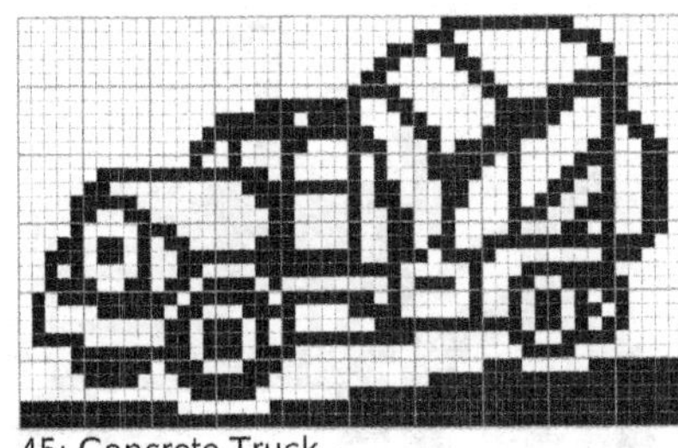
45: Concrete Truck

46: Girl

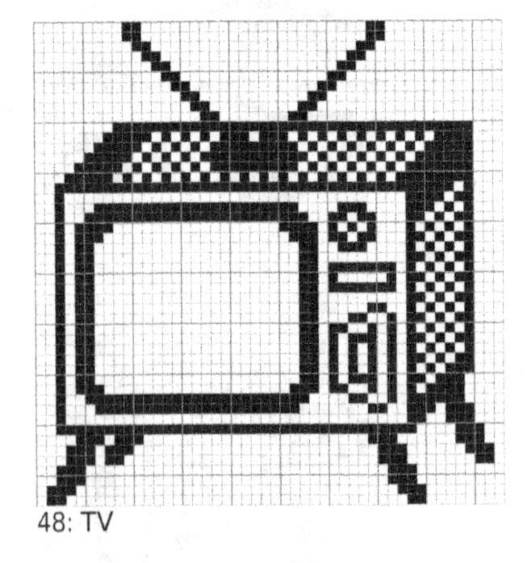
48: TV

49: Dark Magic

47: Photo Camera

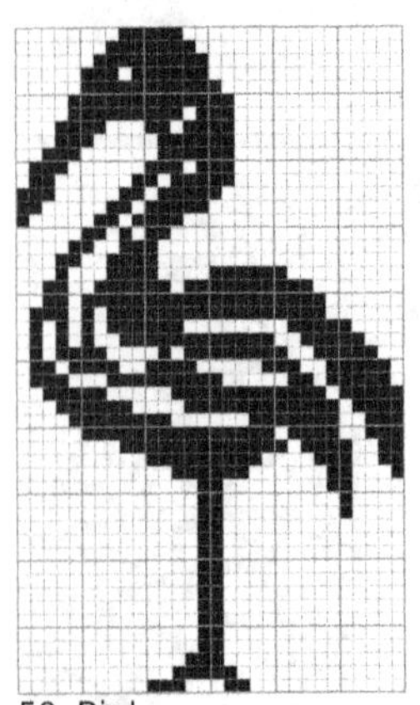
50: Bird

51: Funny Hedgehog

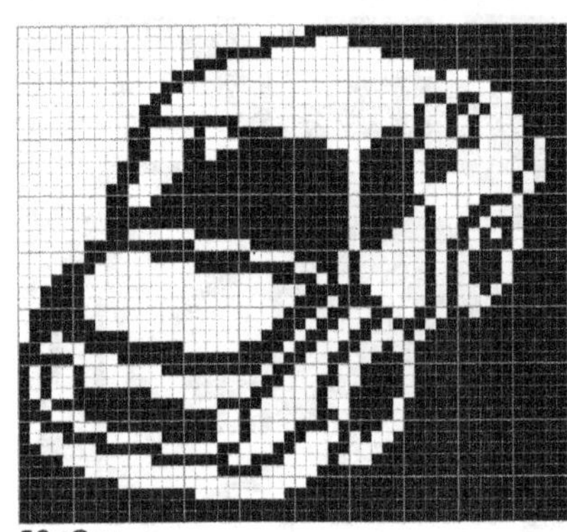
52: Car

53: Cat

54: Thomas!

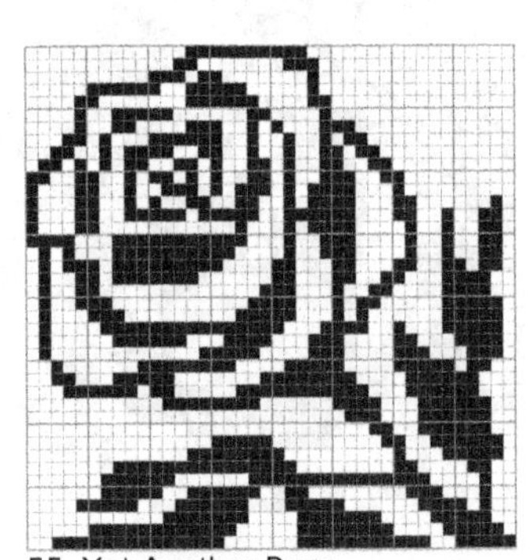
55: Yet Another Rose

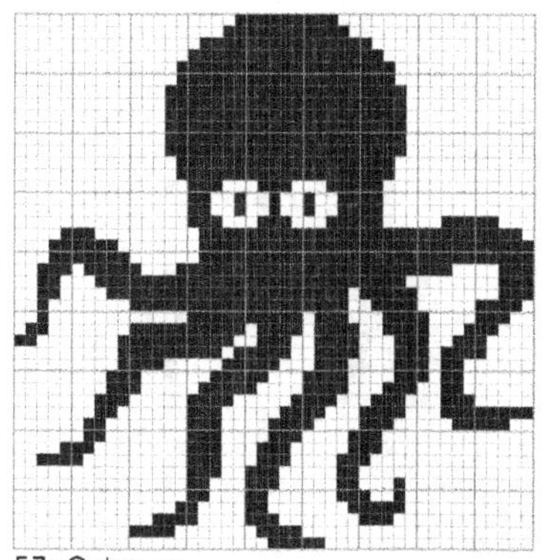
57: Octopus

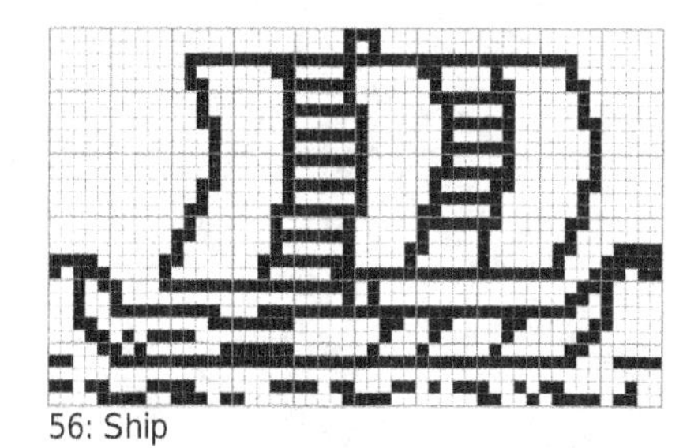
56: Ship

58: Harvester

59: Boy

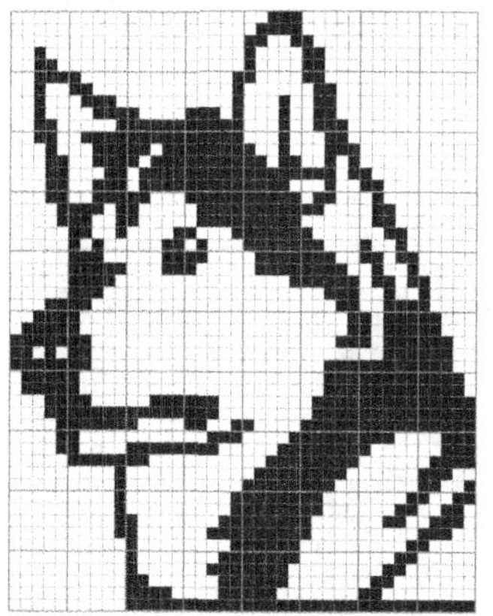
60: Postman

61: Lamb

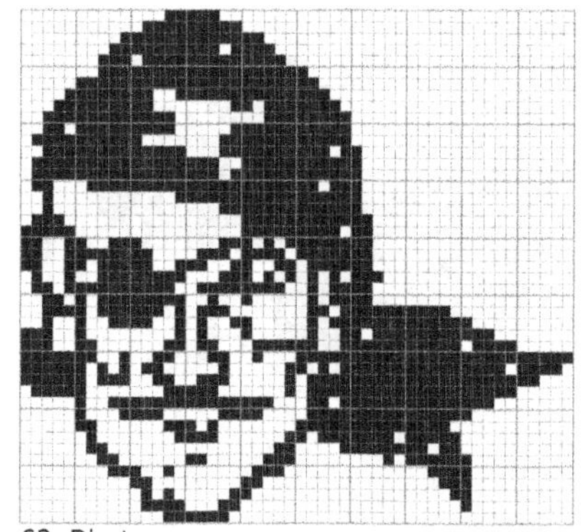
62: Pirate

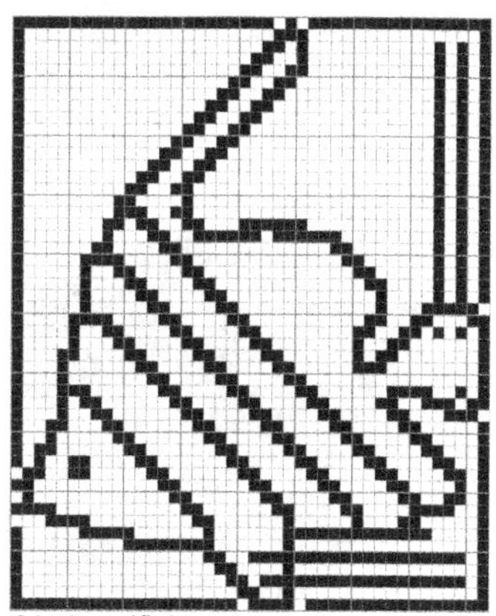
63: Washington Huskies

64: Angelfish

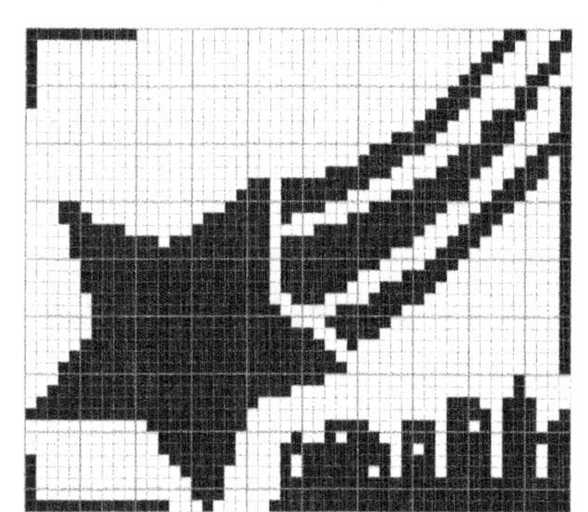
65: Put It in Your Pocket

66: Donald Duck

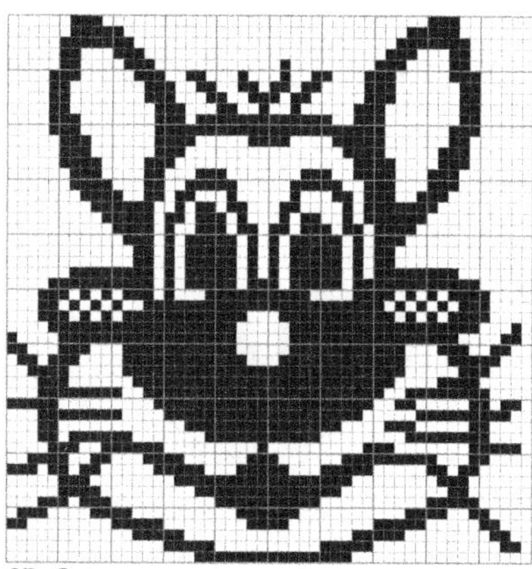
67: Cat

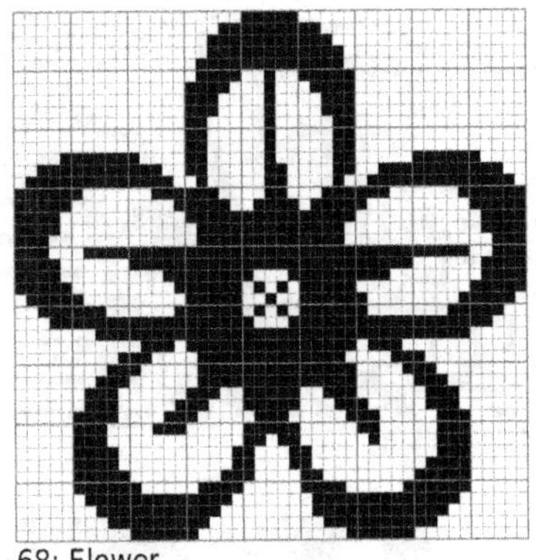
68: Flower

69: Hawk in Black and White

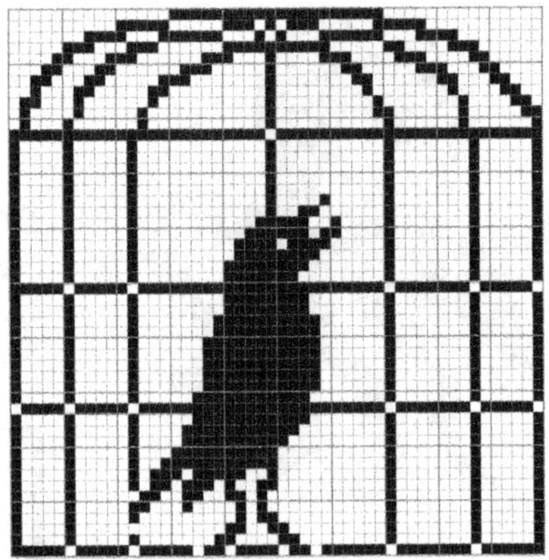
70: Birdcage

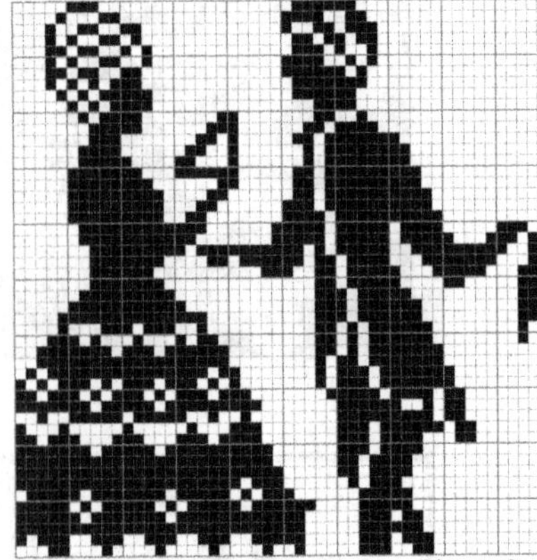
71: Would You Dance with Me, Please?

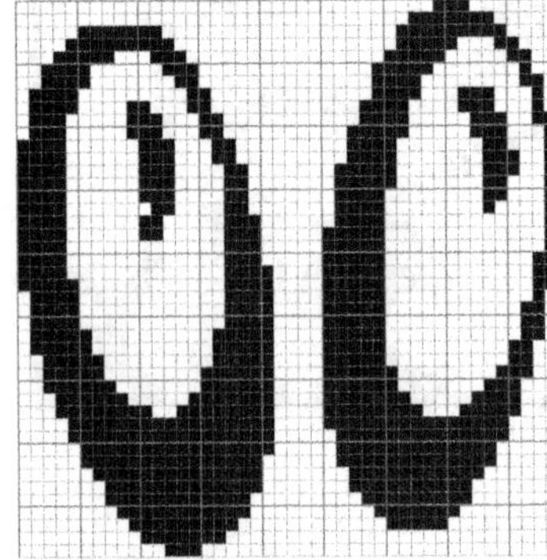
72: Night Sight

73: Lonesome Cowboy

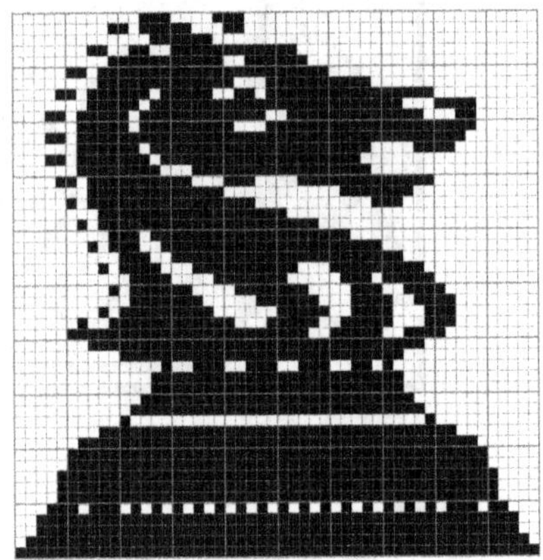
74: Knight

75: Horses

76: Bouquet

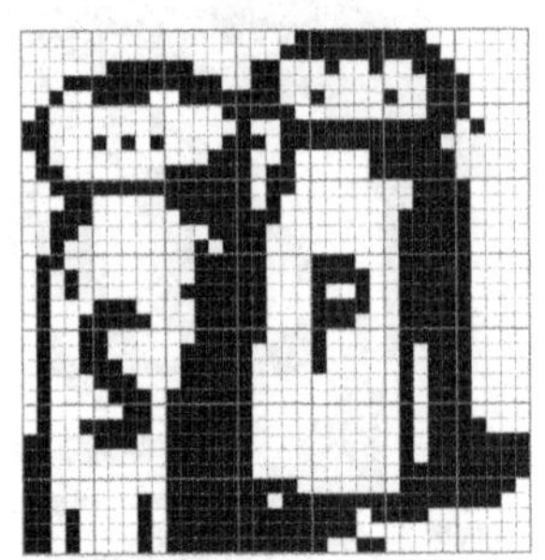

78: Condiments

77: Dance

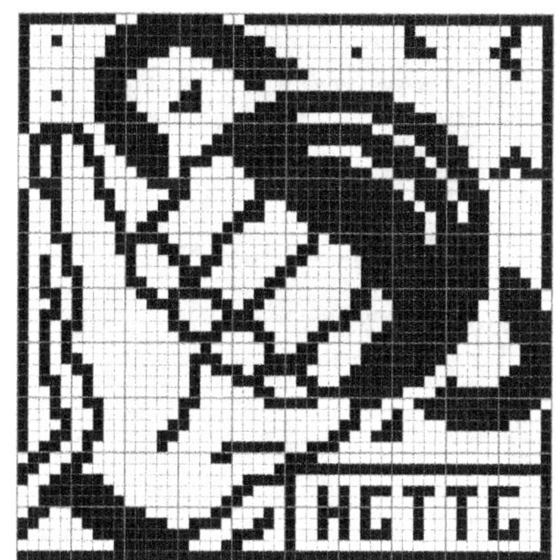

79: Tribute to Douglas

80: Oh Deer

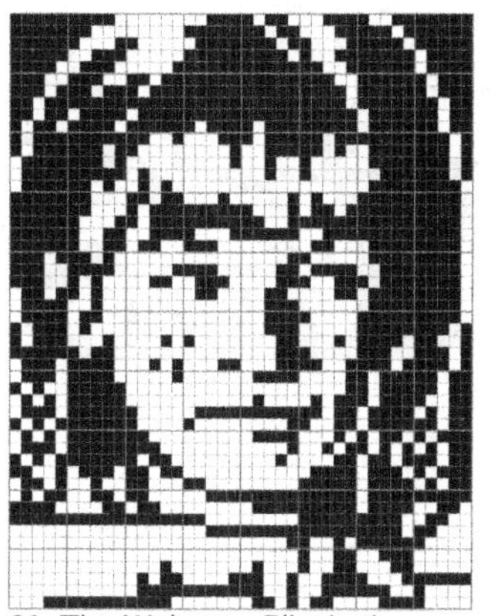

81: The Waltons - Elizabeth

82: Pluto

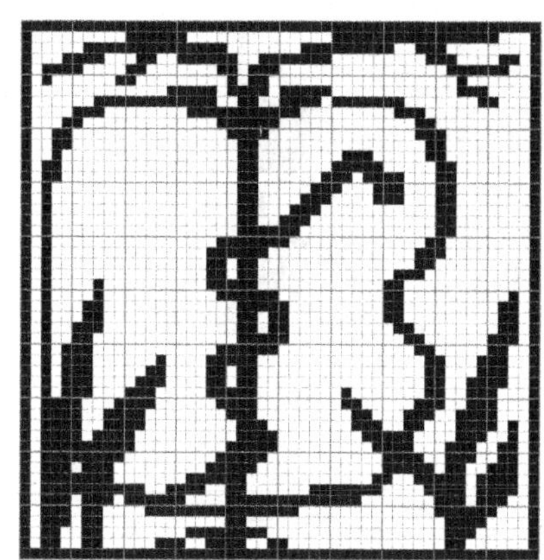

83: Bible Stories

84: Go Away, Kid, You Bother Me!

85: MASH - Hot Lips

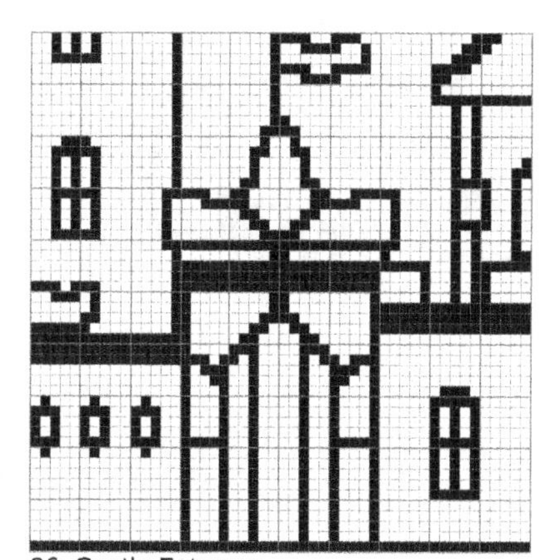

86: Castle Entrance

87: Olive Oil

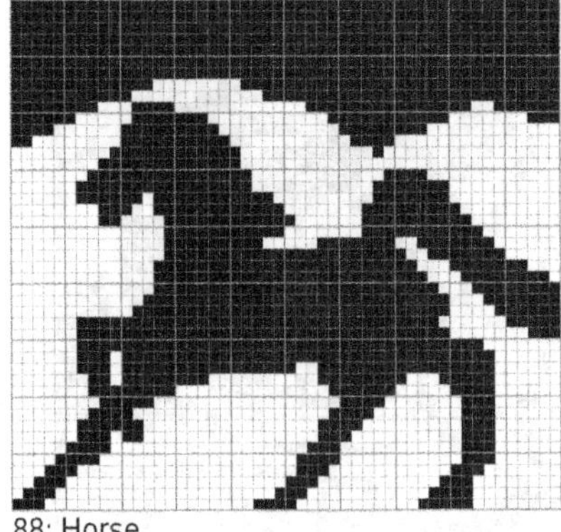

88: Horse

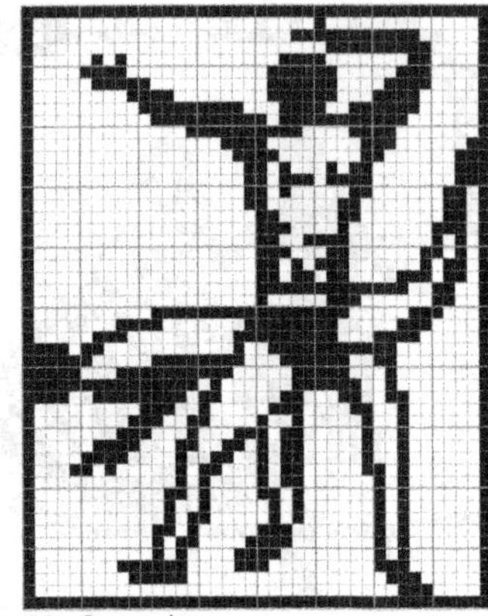

89: Capoeira

90: Chopping Wood

91: Camera

92: Incan Pyramid

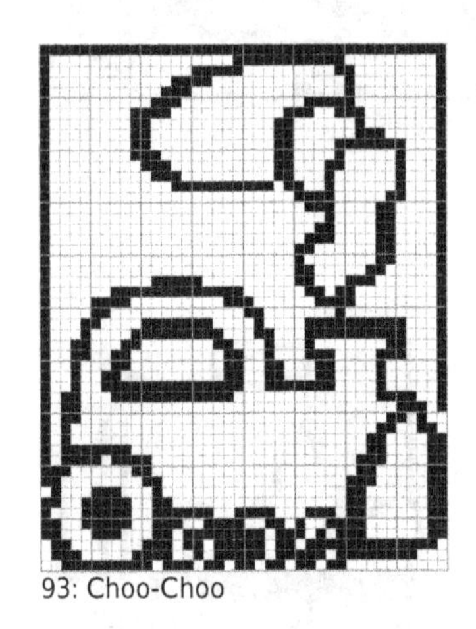

93: Choo-Choo

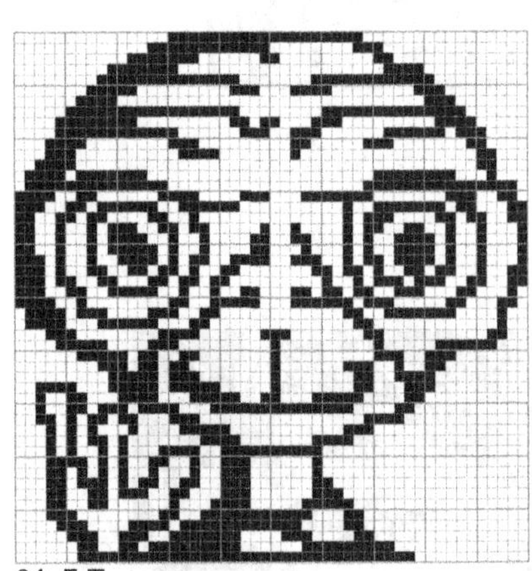

94: E.T.

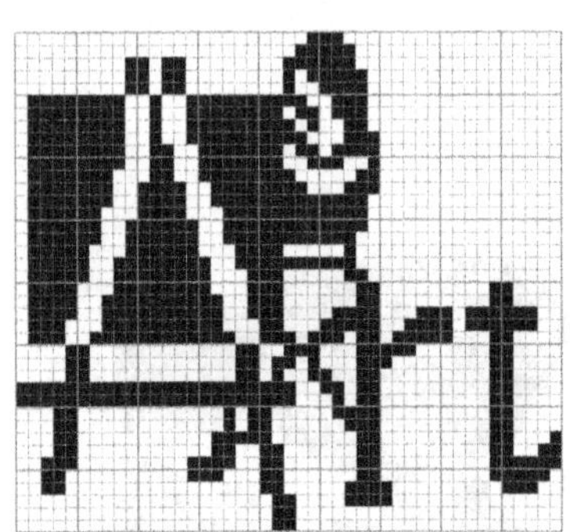

95: Art

96: Birthday Cake

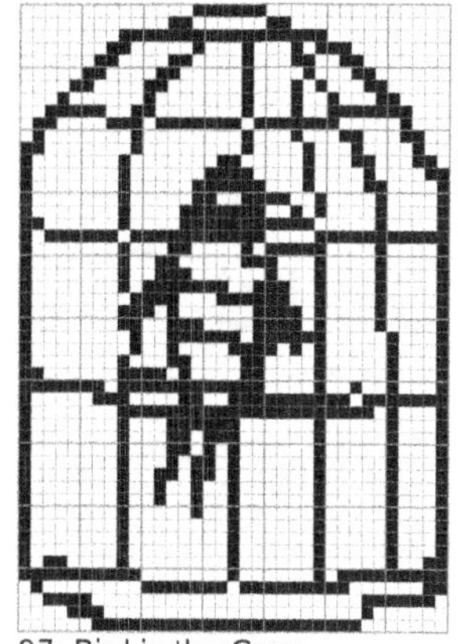
97: Bird in the Cage

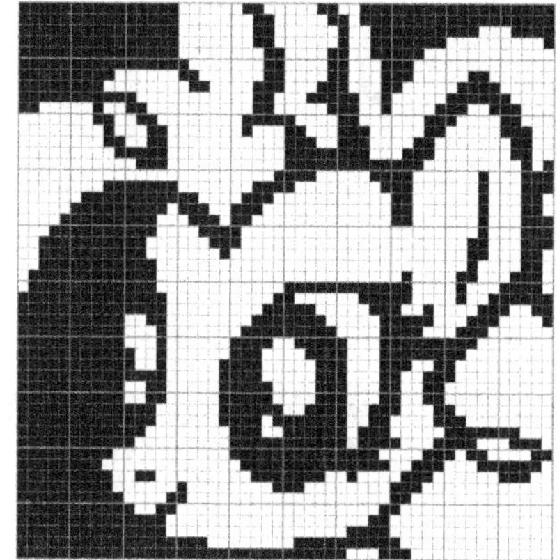
98: Pink is Best

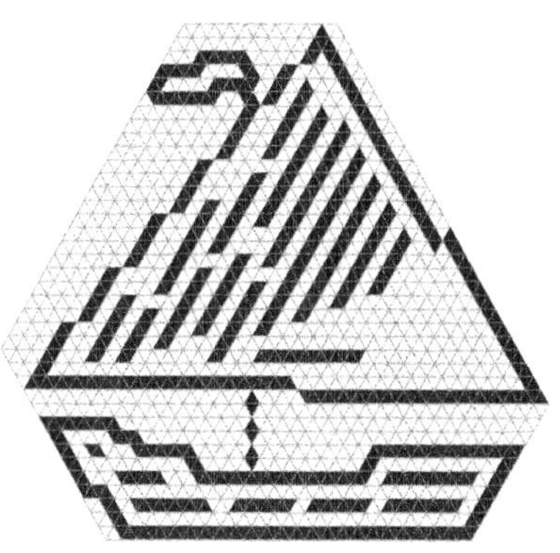
99: Boat

100: Husky

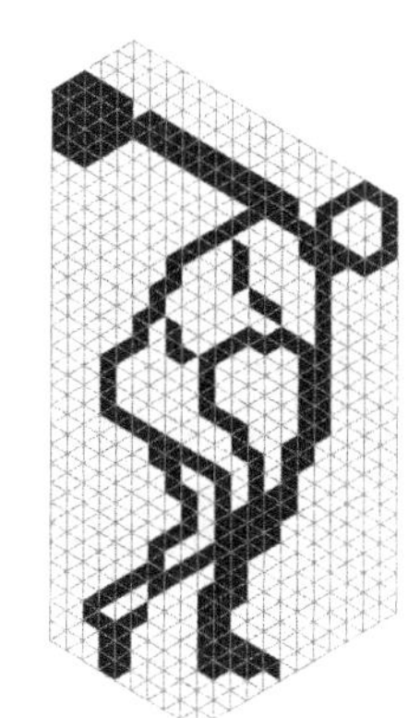
101: Discobolus

104: Horse

102: Monkey

103: Man's Best Friend

GRIDDLERS
Logic Puzzles

Picture Logic Puzzles:

Griddlers

Griddlers are picture logic puzzles in which cells in a grid have to be colored or left blank according to numbers given at the side of the grid to reveal a hidden picture.

Triddlers

Triddlers are logic puzzles, similar to Griddlers, with the same basic rules of solving. In Triddlers the clues encircle the entire grid. The direction of the clues is horizontal, vertical, or diagonal.

MultiGriddlers

MultiGriddlers are large puzzles that consist of several parts of common griddlers. A Multi can have 2 to 100 parts. The parts are bundled and, once completed, create a bigger picture.

Word Search Puzzles:

Word Search

Word Search is a word game that is letters of a word in a grid. The goal of the game is to find and mark all the words hidden inside the grid. The words may appear horizontally, vertically or diagonally, from top to bottom or bottom to top, from left to right or right to left. A list of the hidden words is provided.
Each puzzle has some text and underscores (_ _ _) to indicate missing word(s). If the puzzle was solved successfully, the remaining letters pop up in the grid and the missing words appear in the text.

Number Logic Puzzles:

8	1	2	5	7	9	4	3	6
4	6	9	1	3	2	5	8	7
5	3	7	6	8	4	9	1	2
6	7	4	8	5	1	2	9	3
1	8	3	2	9	7	6	4	5
2	9	5	4	6	3	1	7	8
9	5	8	3	1	6	7	2	4
7	2	6	9	4	8	3	5	1
3	4	1	7	2	5	8	6	9

Sudoku

Sudoku is a logic-based, number-placement puzzle. The goal is to fill a grid with digits so that each column and each row contain the digits only once.

4	2	3	6	5	1
1	5	4	3	6	2
6	1	2	5	3	4
2	3	6	4	1	5
3	4	5	1	2	6
5	6	1	2	4	3

Irregular Blocks (Jigsaws)

Jigsaw puzzle is played the same as Sudoku, except that the grid has Irregular Blocks, also known as cages.

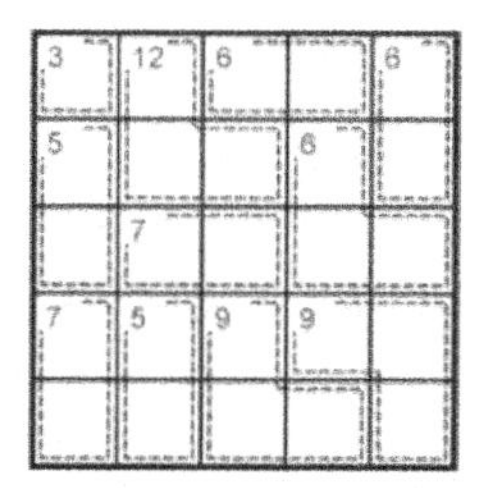

Killer Sudoku

The grid of the **Killer Sudoku** is covered by cages (groups of cells), marked with dotted outlines. Each cage encloses 2 or more cells. The top-left cell is labeled with a cage sum, which is the sum of all solution digits for the cells inside the cage.

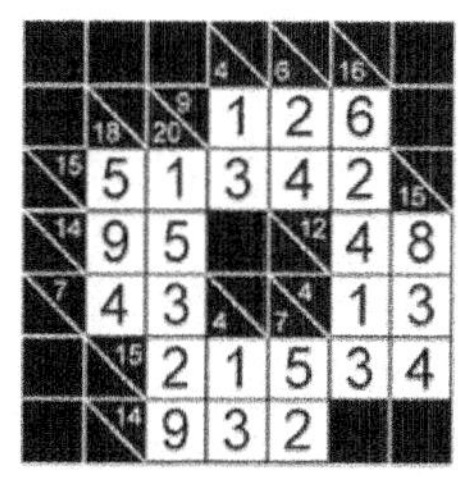

Kakuro

Kakuro is played on a grid of filled and barred cells, "black" and "white" respectively. The grid is divided into "entries" (lines of white cells) by the black cells. The black cells contain a slash from upper-left to lower-right and a number in one or both halves. These numbers are called "clues".

0	0	1	0	1	1	0	1
0	0	1	1	0	1	0	1
1	1	0	0	1	0	1	0
0	1	0	1	0	1	0	1
0	0	1	1	0	0	1	1
1	0	1	0	1	0	1	0
1	1	0	1	0	1	0	0
0	0	1	0	1	0	1	1

Binary

Complete the grid with zeros (0's) and ones (1's) until there are just as many zeros and ones in every row and every column.

griddlers
Logic Puzzles

Number Logic Puzzles:

Greater Than / Less Than

Greater Than (or **Less Than**) Sudoku has no given clues (digits). Instead, there are "Greater Than" (>) or "Less Than" (<) signs between adjacent cells, which signify that the digit in one cell should be greater than or less than another.

Futoshiki

Futoshiki is played on a grid that may show some digits at the start. Additionally, there are "Greater Than" (>) or "Less Than" (<) signs between adjacent cells, which signify that the digit in one cell should be greater than or less than another.

Kalkudoku

The grid of the **Kalkudoku** is divided into heavily outlined cages (groups of cells). The numbers in the cells of each cage must produce a certain "target" number when combined using a specified mathematical operation (either addition, subtraction, multiplication or division).

Straights

Straights (**Str8ts**) is played on a grid that is partially divided by black cells into compartments. Compartments must contain a straight - a set of consecutive numbers - but in any order (for example: 2-1-3-4). There can also be white clues in black cells.

Skyscrapers

The **Skyscrapers** puzzle has numbers along the edge of the grid. Those numbers indicate the number of buildings which you would see from that direction if there was a series of skyscrapers with heights equal the entries in that row or column.

Smart Things Begin With Griddlers.net

CPSIA information can be obtained at www.ICGtesting.com
Printed in the USA
LVOW03s0739290415

436547LV00003B/17/P